ULTIMATELY
EVERY PERSON ON EARTH
WILL CHOOSE TO BELIEVE

In the beginning . . .
God? or Dirt!

Sal Giardina

INCLUDES:
- ➢ 15 ILLUSTRATIONS
- ➢ 3 CHARTS
- ➢ 18 PICTURES
- ➢ 3 TABLES

In The Beginning...God or Dirt?
By
Salvatore Giardina
Copyright © 2010 by Salvatore Giardina

New Creation Ministries, Inc
2025 S. Ash Circle   Mesa, Az 85202
www.GodorDirt.com

Library of Congress control Number 2010934235

Printer in the United States
ISBN 978-0-615-41201-6

While the author has made every effort to provide accurate internet addresses at the time of publication, neither the publisher nor the author assumes any responsibility for errors, or for changes that occur after publication. Furthermore, the publisher does not have any control over and does not assume any responsibility for third party websites or their content.

The ultimate multiple choice test!

In the beginning . . .
God? or Dirt!

(The ONLY two choices – circle one)

-----> (A) The WORLD says *Dirt*

-----> (B) The BIBLE says *GOD*

How can I know for sure ???

Read and think before answering because
This is one test you definitely want to pass!

Author

Sal Giardina - former evolutionary scientist
and registered professional geology consultant

B.S. Mechanical Engineering
M.S. Geology
Air Transport Pilot Instructor
Former USAF Pilot and Astronaut Candidate

Licensed and Ordained Minister
Founder of New Creation Ministries, Inc.

√ out the author's Website:
http://GodorDirt.com

ACKNOWLEDGEMENTS

Original Cover Design by: Mr. Lou Tillotson
Cover Computer Graphics by: Keith Piccolo

Critique of manuscript:
  Dee Rush:
  Richard Giardina

Technical Review:
  Tye Rausch

Foreword:
  Dr. Joseph Kezele

Dedicated to Pam Giardina, my spouse, companion, and closest friend of 37 years. This book was completed through her encouragement and dedication as my wife and mother of our two children.

TABLE OF CONTENTS

Page

INTRODUCTION
The Only Two Choices   1
Eternal God or Eternal Matter?   3
Straight Answers   5
The Big Picture   6

THE ANSWER KEY
Universal Laws of Energy   11
Molecules to Man or Man to Molecules   24
Can Life Really Arise From Dirt?   27
Unimagined Complexity   33

TWO WAYS TO KNOW THE TRUTH...
All Creation – one way to know the truth   37
  The Truth Is Usually Simple   41
   Missing Fossils   41
   Living Fossils   45
   Vestigial Organs   51
   Germs & Antibiotics   52
   Age of the Earth   53
Revelation – another way to know the truth   74
  The Creation Account   83
  The Gap Theory   89
  Compromise by Intimidation   94

ORIGINS- 21st CENTURY UPDATE
Security Zone Briefing   99
  Where Does Life Begin?   100
  What Do You Know About Dune Buggies? 112
  Is There A Society Of Planet Earth? 117
  Probably Yes or Probably No?   120
  Does Intelligence + Information ⟷ Life? 126

The Future of Mankind   139
     Back to the start   144
     Loss of Longevity   146
     Effect of the Flood   155
     Effect of Babel Dispersion   159
     The Genetic Problem   161

NO ROOM FOR COMPROMISE
Biblical Revelation or Human Speculation 171
     The First Humans   175
     Origin of Energy   180
     First Light of Creation   185
It's Time for a Decision   190

CREATION SCIENCE ON THE WEB   195

SUGGESTED ADDITIONAL READING 196

FREE POWER POINT RESOURCES   196

FOREWORD

Today there is so much misinformation and deception regarding the issue of the origin of the universe and of the human race that a fair number of people are confused. Some believe that everything we see is the result of natural processes occurring over very long periods of time in a random, haphazard course of events with no intelligent input, plan or purpose. In this world view there is no hope. Others believe that the God of the Bible created the universe, planet Earth and everything on it including humans in their present form in six ordinary days a little over 6,000 years ago, certainly less than 10,000 years ago. A third group believes that somehow the God of the Bible used evolution over long periods. They attempt to reconcile the conflicts between the Biblical account and the evolutionary scenario.

In the Beginning....God or Dirt lays out the key issues in extraordinarily clear language, so that a high school student and those with higher education can easily understand why they need to consider them, for the results of decisions made (or not) are eternal. Sal Giardina presents the basic laws of science succinctly and plainly, so that it is easy to understand how the evolutionary model runs counter to those laws of science. Evidence in various fields of science easily demonstrates that the creation model best explains what we observe in the world today. Up to date topics, especially current findings in DNA, radiometric dating and the area of Information help the reader to see logically the impossibility of evolution.

In addition, a very important section dealing with the straightforward meaning of Scripture in the Holy Bible, especially the early chapters of Genesis, is particularly relevant for those who claim to be Bible believing Christians who also hold to an old earth and evolution as part of God's creative acts. With their attempt to compromise Scripture in order to accommodate evolution they think they are somehow assisting God in explaining what He did, and that He is not capable of communicating clearly to all people of all ages. They also may be thinking that they can have the respect of the world as well as salvation from the Lord. Sal Giardina does a good job of demonstrating how that person fails in that goal and risks earning the contempt of both man and God.

This book serves to present these most basic and critical concepts from both science and Scripture so that the reader, if he is intellectually honest, will come to the conclusion that we were created by the God of the Bible as He said in His Word, with significant amounts of data from observed evidence confirming Scripture. This leads the unbelieving and compromised readers to their realization for their need to accept the gift of God, salvation through acceptance and belief in Jesus Christ as their Lord, Savior and Master, with evolution having no part whatsoever in actual history.

Joseph Kezele, M.D.,
President, Arizona Origin Science Association

CAUSE - EFFECT
PURPOSE OF THE UNIVERSE

God's Biblical Explanation:

Genesis 1:1 In the beginning God created the heavens and the earth.

Revelation 21:1 And I saw a new heaven and a new earth; for the first heaven and the first earth passed away...

Revelation 22: 1-5 And he showed me a river of the water of life, clear as crystal, coming from the throne of God and the Lamb... And on either side of the river was the tree of life... and there shall no longer be any curse, and the throne of God and the Lamb shall be in it... and they shall see His face...and there shall no longer be any night... because the Lord God shall illuminate them; and they shall reign forever and ever. (NAS)

Humanity's Naturalistic Explanation:

Evolution 101 In the beginning the formless Eternal Elements began to organize into the ordered celestial and organic systems of the physical universe.

Darwin's Origin of the Species Eventually, after eons of time, all life forms developed as a result of random accidents devoid of preordained design.

Human Manifesto The culmination of evolution is the human mind which now contemplates the wonder of this fortuitous series of events. Unfortunately, each life is of limited duration, having no hope of long term plan or purpose.

INTRODUCTION

The Only Two Choices

As you pick up this book, you are probably wondering about the title and mumbling something like this to yourself; "What does 'In the beginning...God or Dirt' mean, anyway?" And what's this all about - "Only Two Choices?" Well, just read on and hold on, because you can expect to receive some marvelous insight into the origin of the universe. In addition, you will be provided with tools which will enable you to see through the misinformation which has caused many people to neglect or reject the truth of God's Word (the Bible) concerning creation. This deception has not only influenced multitudes of non-Christians but, unfortunately, has also crept into many churches due to a lack of understanding of the relationship of science and the scriptures.

Although the truth is relatively simple, it has not always been a quick nor easy task to assure oneself that all the scientific disciplines point with the precision of a laser toward an intelligent designer. As you read this book, however, it will quickly become obvious that there exists many irreconcilable contradictions concerning the current popular explanations for the origin of life from dirt (non-living matter). Even a casual reading will expose you to an updated summary of comprehensive evidence that is consistent with the biblical record of creation and the global flood. This book will bring you right up to date and transport the reader into the light now made available by the Information Age; however you should recognize that the Creator of this universe permits **all** people to make their **own** decision. This magnanimous gift of sovereign choice elevates man far above any animal and is based solely upon God's infinite love for mankind.

Understanding origins therefore, can clarify each individual's perception of God's character and also helps answer the universal question- "Why is the world in its present physical and spiritual condition?" Could the answer be that most of mankind has been making the *wrong choice* since Day One? Now, the straight forward information presented in this book can help you to navigate through the smog of unfounded evolutionary theories by providing sufficient reason to believe that:

"In the beginning..."

should be followed by **"GOD"** and not really by **"DIRT"**

I call heaven and earth to witness against you today,
that I have set before you life and death, the blessing
and the curse. So CHOOSE life in order that you may
live, you and your descendants. **Deuteronomy 30:19** (NAS)

Eternal God or Eternal Matter?

Well Sal, what do you mean by 'dirt' and all that? As you are probably aware multitudes of people from every nation have been convinced that the Genesis account of creation should no longer be taken seriously. The deception is packaged and labeled as science and sounds like this: "...the Bible is not compatible with modern science because evolution has proved that mankind was not created. Life merely developed, without plan or purpose, from **non-living chemicals (dirt)** by natural processes as a result of random chance or accident over countless millions of years."

I'm sure you have heard this scenario many times. Perhaps as far back as you can remember. And today every possible form of media is being used to promote this concept in the minds of men. Is it possible that you heard this story so often that you may have even believed it? I did - for a while! For over one hundred years generations have been told **AS FACT** - not theory - that the Bible should read **"In the beginning...DIRT."** Why so? - because it is maintained that the entire universe started out with only unformed matter, perhaps hydrogen atoms. It is then *reasoned* that the only god needed to create this marvelous universe is TIME itself. You are assured, by pipe-clutching PH.D.'s, that given enough TIME, all things are possible. Therefore, you are asked to believe that a hypothesized singular event, popularly called the BIG BANG, apparently was sufficient to cause hydrogen gas to ultimately develop into the ordered physical universe as well as the most sophisticated systems imaginable - YOU & ME!

Has anyone ever seen order result from
the chaos of an explosion?

Please, don't laugh. Even though such conjecture is known to defy mathematical probability and many established laws of physics, multitudes of people have bought this lie of the **Master Deceiver.** The deception, you see, is consistently concealed in an official looking report with a return address label as if it was the actual product of an objective "research lab." Indeed the theory of evolution has been endorsed for decades by a great portion of the scientific community, having been marketed and packaged by the popular media, school texts and museums to sound and look scientific. But what you need to understand about Genesis and Creation is this:

IT TAKES JUST AS MUCH FAITH TO BELIEVE
In the beginning ...DIRT
as
In the beginning...GOD

Straight Answers

This book has been prepared for people who may not have the time or background to wade through the mountain of evidence or evaluate the numerous debates concerning origins. For this reason, it was not the author's intent to provide an exhaustive scientific treatise on the subject. The information is even sometimes condensed to brief thought provoking statements in order to get right to the focal point of an issue. However, ample references are cited for those who desire to verify the factual content.

No doubt, you will enjoy learning why the most fundamental scientific principles have direct application to the understanding of the origins of life. The information presented should provide logical answers for any open minded person who desires to settle this intriguing question:

Was life started by God through creation

or

from Dirt through evolution ??

If for whatever reason you may have believed the latter, be prepared for the necessity of some mental readjustments long before the last page of this book is read.

The Big Picture

Did you know that there are many aspects of evolution that science can no longer justify? If you are puzzled by such a question it is because this state of affairs is not promoted by the popular media nor may it be mentioned in most schools due to intimidation of the faculty. Even those scientists who choose to maintain a belief in naturalism must inevitably come to a point of inconsistency in the face of available evidence. At that point, one's belief in evolution becomes a self-willed philosophy maintained solely by faith in the religion of Atheism.

You have probably heard that the Information Age has provided mankind with more knowledge in the last several decades than has accumulated over many centuries. This is indeed true, but is your personal information base outdated simply because you have not kept up with the pace? Are you aware that currently there is readily available more than sufficient data to overwhelmingly resolve the age-old question posed on the cover.

The long standing idea of life resulting from chance and accident, in light of today's discoveries, is no longer justifiable or even rational. The research of microbiology alone has documented that a living cell is in reality an extremely small nano machine way beyond the complexity of any product created by human ingenuity. Scientists currently do not have a comprehensive understanding of the myriad of internal operations of cellular life nor is there a satisfactory evolutionary explanation for the interdependence required between the vast ecosystems of our planet.

The information age has, however, provisioned us with clear insights into the magnitude of the complexity of life. So now, with the literal explosion of this revelation, the inadequacy of neo-Darwinian theory has become evident. Yet why would thinking people continue to embrace a theory that has minimal supporting evidence at best, and a plethora of current high tech research to negate it? Perhaps life is just moving by too fast for many to slow down and think about it. On the other hand, I believe that a considerable number of souls refuse to acknowledge God, the creator, simply because they do not want to be accountable to Him, the God of love who permits every person to make that choice.

As you read this book you will shortly come across the extremely important principle of:

The 2nd Law of Thermodynamics.

Don't panic - you do not have to be a mechanical engineer or person trained in the applied earth sciences to understand the application of this law. Actually you live with it every day. It is emphasized, however, because it is acknowledged as a universal law of the physical sciences with no known exceptions. For example, any process, machine or living system that is proposed to ultimately violate the universal tendency of deterioration, running down, decay and decreasing order which is governed by the 2nd Law would rightly be immediately rejected as impossible.

All engineers and physicists know there can be no perpetual motion machines in the presently existing physical universe. However, the proposed process of evolution essentially requires a construction project analogous to building a biologic perpetual motion machine. In fact, the 2nd Law is so well established that it stands as a monolith, permitting all mankind from farmers to rocket scientists to invent and construct everything from widgets to nuclear reactors with the certainty of predicting their long term limitations.

Why then, has the idea of organic evolution been so widely accepted when it proposes a process which yields ever increasing order and complexity which is directly opposed to the 2nd Law? The theory, in reality, is not unlike many misguided inventors who attempt to patent a perpetual motion machine. All that has ever been observed, and observation is the basis of science, is a continuous breakdown of every product of the natural and living world into less ordered and complex chemicals. This is why our cars wear out and our beaches are covered with sand. This reality only confirms the 2nd Law and stands opposed to the hypothetical conjectures of naturalism. By the way– you will not find a perpetual motion machine anywhere in the world.

Is Any Of This Relevant To Me?

Much of one's life direction is indeed governed by the choice of answer held within the inner recesses of your heart.

- Did life spontaneously appear and then continue on a long journey of one creature changing into another?

  If so, there is really no need for a creator. Thus every man is authorized to live as he thinks is best for himself. After all, natural selection will eventually eliminate those who are falling behind. The world has no ultimate meaning and the value of human life is greatly diminished.

- On the other hand, can the physical world and life be perceived as part of a Divine plan that includes purpose and a long term relationship with its creator?

  Your life in the here and now as well as the hereafter could depend on the correct answer!

Perhaps you are not an introspective person but are more of an analytical thinker. So you are probably asking yourself this type of question. Is it really possible to reject evolution from a rational or intellectual standpoint? This is a valid question. And thus, the purpose of this book is to affirm that abundant evidence is available to allow the average person on the street to answer this question without hesitation.

"Come now, and let us reason together"

Isiah 1:18 *(NAS)*

THE ANSWER KEY

Universal Laws Of Energy

You see, the average person on the street is not a scientist. Therefore, regardless of his education he may not know the following: The total current data base of all scientific information provides **No Theory,** or **No Explanation,** or **No Evidence** – not even a **Clue** as to where the original **DIRT** came from! Herein science falls short, being totally bankrupt and mute. Recently, I heard a scientist defending evolution by saying "there is no reason to conclude that matter was ever created." In other words he was saying the best scientists can offer is the **assumption** that matter has been around forever! This mega-assumption obviously requires a great leap of faith.

Now if you are inclined to take this leap of faith, please understand that science cannot come to your rescue and back you up. On the contrary, our current knowledge of scientific principles actually indicates that matter has not been hanging around forever. This leaves you at a place where your faith in the eternal existence of matter is no longer intellectually justifiable in light of the current knowledge of scientific laws.

Notice that I said "Laws", not theories or philosophies. Scientific Laws like Newton's Law of Gravitation have no known exceptions. Jump off any 20 story building and it's all over! The law of gravity never fails. Have you ever wondered how we can send astronauts to the Moon and

Scientific Laws are discovered by:

Observation
Experimentation ⟺ **Known Systems**

Application of Scientific Laws are always:

Predictable ⟺ *Repeatable*

and therefore
Operate Without Exceptions

The absolute basis of refuting a….

MODEL | IDEA | THEORY or

SPECULATION ….. is simply by demonstrating

that it violates a mathematical, chemical, or physical Law of Science!

return them safely? These endeavors are only possible because of the **reality** of physical **laws** that permit predictable consistent engineered applications. Likewise, the First and Second Laws of Thermodynamics have been established as incontrovertible for hundreds of years. Now just relax, you do not have to be a physics major to understand what I am about to show to you. But after I explain this, neither will you have any scientific grounds for believing in eternal matter.

THE FIRST LAW: You may have heard of it—

Energy or matter can not be created or destroyed
(in the physical universe)
but is only transformed from one form to another

This law leads to the direct unalterable conclusion that all the matter and energy within the universe is a fixed quantity - remains constant. This law has been consistently verified over centuries and has no known exceptions. Now even though the total quantity of energy/matter in the universe never increases or decreases, we can also count on another universally accepted principal —

THE SECOND LAW:

> The amount of energy available to do work in the universe is always decreasing.

Like the First Law, the Second Law is fundamental to science and is used routinely by engineers in the design of many machines. This law actually gives us great insight into the origin of the physical universe. Hang in there and follow this explanation.

Basically we can say for certain that even though the total quantity of energy in the universe is fixed, the character of this energy is continuously **degrading** from available useful energy to an unavailable, unusable form. For example, every time a machine operates or any energy conversion process occurs, a certain amount of useful energy is dissipated as unusable low temperature heat. This is why perpetual motion machines are not possible. The dissipated energy remains in the universe (none is ever lost) but is unrecoverable for any further useful work.

Every person experiences the application of the 1st and 2nd Law throughout the day without being aware of this interaction as they go about many common routines. The use of a toaster or any small household appliance always involves the operation of these principles. For example, whenever the brakes of your car are applied you are converting the mechanical (kinetic) energy of the moving vehicle to heat energy which is dissipated into the

14

atmosphere as unrecoverable energy. Although this energy is converted from one form to another without loss, the total energy involved is conserved (1st Law) but it is no longer available to be recovered for any other application because it has been dissipated into the cosmos (2nd Law).

Knowing this allows us to understand that the age of the universe can only go back so far to a point in time when all the initial energy had to be in the form of useful available energy. This point in time would then be 'The Beginning' because the universe could not be any older. It may be helpful to understand this concept by relating it to a beautiful Grandfather clock that you may remember. Kids always ask what makes it run so long **by itself**? Well didn't you answer by saying 'someone had to wind up the clock– In the beginning'! Now we can definitely reason that the clock never has nor can it wind itself up, nor can it run forever by itself! This means that it cannot create or supply its own energy (1st Law). Now just as the current **operation** of the clock can not be infinitely old for it had to be initially wound up, so also it follows that the universe had to be supplied its initial useful energy from a source beyond (outside) and independent of the physical universe in accordance with the Laws of thermodynamics.

Yes - but just what does all this mean anyway? Just that it reinforces the Bible's account of origins. Namely, these two universally acknowledged laws shout that the physical universe absolutely requires a beginning. The **"DIRT"** could not possibly be infinitely old. The First Law commands that the physical universe is a closed system with a fixed amount of total matter $\leftrightarrow$ energy, which can never increase or decrease. However, the Second Law reveals to us that all the useful energy available, that which man can use for practical applications, is less than the total because it has always been diminishing. (refer to illustrations and text at the end of this section)

The 2nd Law is very significant in that it actually allows us to look back in time. We can therefore conclude with assurance that in the past, the amount of energy which was available to perform useful work was greater than today. Thus if the universe were in fact infinitely old, then at some point in time in the dateless past, the amount of energy available to perform work would eventually exceed the total energy in the universe which according to the First Law is fixed. This is impossible so the only reason-able conclusion is that the mass of the universe (dirt) is not infinitely old but had to have a beginning. This conclusion is in perfect accordance with the account of creation given to us in the Bible.

If this seems a bit confusing, you may also look at this picture from another perspective. Eventually, the 2nd Law requires that the universe will reach a stable state whereby it will no longer have any useful available energy remaining. At that point in time all energy processes will cease. This final state has not yet been reached; therefore, the matter which comprises the universe cannot be infinitely old. This information leaves us with the **objective facts** of science requiring a beginning for the physical universe. Going further, these two laws clearly attest to mankind that the universe had to begin as a highly ordered system that has **deteriorated** with time to the present state. This tendency whereby everything trends 'downhill' into disorder is foundational to the sciences of physics and chemistry, being designated as the inevitable constant movement toward increasing entropy. Entropy is used by engineers as a measure of the disorder in a system. However the concept of entropy has apparently been announced in the Bible for millennia (Genesis 3) long before our understanding of these natural laws. How then, you may ask, could the present state of the universe consisting of order and complexity be assembled from the random chaos of the "Big Bang"? A reality check shows us that the current state of knowledge regarding the physical properties of the observed universe does not point to a catastrophic origin. On the contrary, the established laws of science point to an original ordered universe of finite age that is continuously becoming more disordered.

Where does all this leave us?

Because there are no other alternatives possible, ultimately each individual must choose whether he will direct his life according to:

Ⓐ - the world's popular concept of origins that is not supported by the most fundamental universal Laws

or by:

17

Ⓑ - the clear teaching of scripture which is supported by the fundamental principles of science from which mankind has consistently reaped so many technological benefits.

So the bottom line boils down to this. **Every person** will exercise his or her faith in one of the following beliefs:

ETERNAL GOD

or

ETERNAL MATTER

But now you have been informed…

> The established laws of science preclude eternal matter and the evolutionary concept of perpetual movement toward a state of increasing order!

For those who choose to believe that matter has existed eternally with no explanation, regardless of the facts, please be aware that there is also an additional step of faith required. You see, you may not have been told that there has NEVER BEEN a mechanism ever **demonstrated** by science by which non-living chemicals can assemble themselves into complex life forms. This statement is supported by the incontrovertible observations through out all of human history that further support the validity of the 2nd Law. Namely, that all objects, chemicals, or systems which are left to themselves always become less ordered and eventually break down into simple molecules. This tendency has been a nuisance for mankind as it is common knowledge that **everything** ultimately wears out and deteriorates into its component parts as time progresses. Nothing devised by man has ever been able to stop this process. At best, man with all his technology can sometimes slow the process down a bit and prolong the inevitable. As an example, consider the modern sophisticated aircraft. A commercial jet aircraft is probably one of the most care-fully maintained systems in the world. But think about it - even with **man's** persistent technological intervention of main-tenance and refurbishing, still in time the system (aircraft) wears out and must be replaced. And of course, not even human life can be prolonged indefinitely. The bottom line is this:

> No scientist has ever demonstrated by an acceptable scientific method the opposite effect, whereby matter can construct itself into more complex chemicals or systems under natural conditions.

(inorganic molecules or systems)

The undisputed long term results of natural events

The theory, commonly known as evolution, requires the elements of the universe to order themselves into more complex systems or organisms under natural conditions.

This infers the total absence of intelligent intervention. The idea appears to be reasonable when it is related to organic living systems but this is merely an illusion. Think about it, is there anything you know of that escapes the second law? (organic molecules or systems)

The undisputed long term results of natural events

In light of this immovable stumbling block, most evolutionary scientists, therefore, are not eager to point out that the proposed assembly process called evolution also runs completely opposite to the known state of affairs demanded by the Second Law of Thermodynamics. This violation, in reality, renders evolution theory to the realm of *science fiction*. Recall that all that is needed for the absolute refutation of a theory is to demonstrate that it violates an established Law of Science!

So again a thinking person is faced with only two alternatives…

<u>**An Eternal God created the universe and all life in it**</u>

OR

<u>**Eternal Dirt eventually constructed *itself by accident***</u>
<u>**into the complex**</u>
<u>**ordered physical universe and all its life forms**</u>

So then, from an academic standpoint the book should have already been closed on the concept of physical and biologic evolution. However, because not everyone is a scientist, a more complete understanding of these fundamental principles will be related to recent discoveries. Perhaps what follows can help you to make YOUR OWN choice.

WHAT DOES SCIENCE REALLY SHOW US?

**Answer: Matter/Energy cannot create itself!
because**
↓

The 1st Law of Thermodynamics permits no exceptions.

It is called the Law of Conservation of Matter & Energy

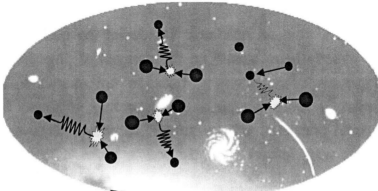

Physical Universe

NOTE: The physical universe is not infinite. **Psalm 147:4**

● *Matter* ⟶ *Motion*   ⟿ *Energy*

➤ There exists a fixed amount of matter/energy in the universe

➤ The total amount of matter + energy is constant (fixed)

➤ Mass converts to energy and Energy converts to Mass
 but the sum of Mass + Energy always remains constant

Where then did the Mass/Energy originate from?

⇨ *SCIENCE HAS NO EXPLANATION* ⇦

THE BIBLE DOES!

GENESIS 1:1 ....
 "In the beginning God created TIME, Space & Matter..."

22

WHAT DOES SCIENCE REALLY SHOW US?
Answer: The physical universe cannot be infinitely old!
because
↓

The 2nd Law of Thermodynamics indicates that the energy
available to do work has always been decreasing!
It is called the Law of Entropy

check out this illustrated principal

➢The universe is limited to a fixed amount of energy
(illustrated as the total area within the circles)

➢At no time in the past could the energy available to do work have
exceeded the TOTAL fixed energy of the universe
therefore

The universe had to have a beginning!

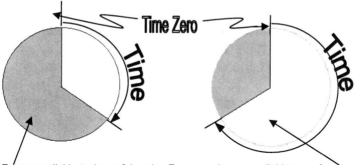

Time Zero

Time   Time

Energy available to do useful work   Energy no longer available to perform useful
work

➢ If matter was eternal, then the universe would be infinitely old,
leaving no energy remaining to perform any useful work.

*The Bible gives us the only source of explanation
for the beginning...*
⇨ CREATION ⇦

Molecules To Man   or   Man To Molecules?

At this point you have examined the illustrations and read the brief text concerning this fascinating subject of origins. You may also be saying to yourself, "things just can't be this simple. Surely the multitude of evolutionary scientists must know how order can be produced from disorder. There must exist documented exceptions to the 2nd Law that are scientifically verifiable". The fact is you may make an exhaustive literature search or research the Web, but you will not find any exceptions which will permit a scenario like that shown in Figure 1. The figure provides a visual representation of proposed evolutionary development. This scenario includes the ordered arrangement of the entire physical universe as well as all life forms.

The laws of thermodynamics and their implications for origins have already been scrutinized by many scientists. Even using rigorous mathematical equations, the results always indicate that complex systems, including living organisms, cannot arise by random chance over time. The universal tendency toward disorder (Figure 2) simply will not permit it any more than it will allow a perpetual motion machine. Figure 2, on the other hand, represents the 'state of affairs' which actually exists today and has been experienced throughout all of human history. Obviously then, the ideas of evolution are completely out of phase with observed reality.

In the recent past, a few evolutionary scientists have proposed that living organisms represent open systems that receive energy from the Sun. It was reasoned that this energy input could, in itself, yield a condition of random exception and **momentarily** organize molecules into more complex systems. However, many other scientists have subsequently pointed out that energy, by itself, cannot produce order in either animate or inanimate material in violation of the 2nd Law. Even, if hypothetically, a short term ordered condition could be produced by chance, the overwhelming and relentless tendency of the entropy principal, as illustrated in Figure 2, will ultimately prevail.

Both mathematics and chemistry research demonstrate that in order for the impartation of energy to yield an increase in order and assemble a working system, an external mechanism and source of information is always required to direct the process. Thus a **pre-existing** machine or intelligent manipulation is needed right from "the beginning".[1] The result of undirected raw energy yields either rapid destruction or ultimately produces disorder and disintegration over time. (Figure 2)

The consensus of scientific observation corresponds to the facts of reality and therefore retains the conclusion …

The Laws Of Thermodynamics Stand Without Exception

REFERENCES:

1. Thermodynamics And The Development Of Order, edited by E. Williams, Creation Research Soc.,1981

Evolution Theory
Vs
Universal Trend of Matter

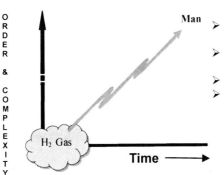

Figure 1

➢ Evolution followed by integration (molecules to man)
➢ Random accidental assembly yielding increasing order
➢ Never observed occurring
➢ Theory:
  Lacking a viable mechanism
  Contrary to known Laws

Evolution 101: " Given enough time all things are possible"

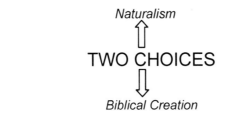

Naturalism

⇧

TWO CHOICES

⇩

Biblical Creation

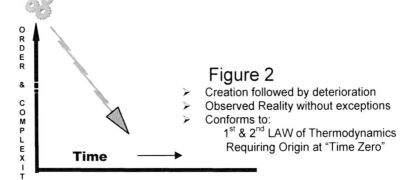

Figure 2

➢ Creation followed by deterioration
➢ Observed Reality without exceptions
➢ Conforms to:
  1st & 2nd LAW of Thermodynamics
  Requiring Origin at "Time Zero"

Genesis 1:1:
All living & non-living systems created by God "In the beginning..."

Can Life Really Arise From Dirt?

Making a correct choice may require some additional knowledge about the complexity of life. The theory of abiogenesis proposes that life has started from non-life, or more specifically from non-living atoms and molecules. Popular media, high school and university text books often present various scenarios as if this idea was empirical fact. Basically the picture that is presented starts out with the origin of life starting in a primordial sea or pond. Inorganic chemicals link up to form **self replicating** organic molecules which are then **envisioned** to accidentally organize into the first life sustaining single cell protozoan or bacterium. From this one event, eventually all other life forms evolved on the earth, each one in succession having a greater magnitude of order and complexity.

Stand back now, take a breath and ask yourself "how much scientific evidence actually exists to support this view"? After all, the **corner stone** of your entire life's perspective is influenced by how much faith you have in this scenario. Can the chemical properties of matter along with random combinations over time cause such a hypothesis to be viable, assuming "dirt" has been around for billions of years? Unfortunately, multitudes of people worldwide as well as our nation's student population have been subjected to a one sided affirmative answer to this question. But perhaps the naturalistic explanation of life's origins would not be so readily accepted if it was generally known that a significant number of evolutionary scientists now acknowledge that **the**

link between chemicals and life is still missing. It may seem reasonable that the missing link will eventually be discovered but this perception is misleading. The fact is all knowledgeable scientists understand that the required sequence of events needed to bridge chemicals to life requires a multiple link chain in which **all** the links are still missing. The analogy of each missing link actually refers to the lamentable absence of any known chemical, biologic or genetic mechanisms.

In the 1950's a number of scientists began extensive research in an effort to demonstrate how this great leap of abiogenesis (life from non-life) could occur. Initially experiments like those conducted by Urey & Miller at the University of Chicago appeared to have yielded the synthesis of some organic chemicals under synthetic conditions that were prepared to mimic the **hypothetical** early environment of the Earth. It is now acknowledged among scientists that the experimental conditions were also apparently carefully predetermined to favor the outcome. This intervention of human design in itself obviously can not be considered random occurrence, thereby negating the validity of the experiment. Nevertheless, further analysis of the data did not uncover any compelling information that would support the theory or discovery of a mechanism for the spontaneous generation of life. On the contrary, similar endeavors coupled with the exploding advances in molecular biology are actually producing overwhelming evidence for the **improbability** of the naturalistic origin of life. Yet in spite of these continuous revelations, many scientists refuse to ac-knowledge the implications of the invalidation of the naturalistic explanation of life. If anything at all was accomplished, the flurry of experiments and research exposed problems so serious that the majority of evolutionists now tend to ignore the whole subject of abiogenesis.

If non-life could bridge the gap to a living organism in one chemical jump (link), then perhaps a person could conclude that the link is yet to be found. However, numerous microbiologists along with organic chemists now assure us that the assemblies required to construct this bridge would number in the thousands. Basically then, modern research has greatly **enlarged the gap** between molecules and life and has now exposed the faulty simplistic foundation of evolutionary biology.

> NO NATURAL MECHANISMS
> OR CHEMICAL PROCESSES ARE KNOWN
> WHICH CAN BUILD A PROTOZOAN
> FROM THE WORLD OF CHEMICALS

This conclusion is justifiably logical, particularly in light of the scientific realization that the following minimum assembly process would have to be accomplished by random accidents in order to build the first living cell:

1. Simple molecules must evolve into complex molecules

2. Complex molecules must evolve into simple organic molecules

3. Simple organic molecules must evolve into complex organic molecules

4. Complex organic molecules must evolve into data storage, self replicating molecules such as DNA

5. All these molecules must somehow combine and organize **simultaneously** into a self sustaining living cell which must assimilate and metabolize food, respirate, grow, respond to stimuli and reproduce.[1]

It should also be noted that in the world of chemistry it has been repeatedly pointed out that the synthesis of these early molecules would have to be accomplished contrary to the **known laws** governing organic chemical combinations.

In light of the above, it may now be obvious to any thinking person that the idea of evolution is essentially a philosophy based on speculation, rather than conforming to the standards of a credible science.

Now consider this…

Each person is obviously at liberty to make his own choice concerning what he is willing to believe about origins. But now you are hereby informed that **NO ONE** has ever demonstrated the *reality* of any of these stages in a laboratory. Neither has any scientist been able to propose even a **plausible theoretical blueprint** for any of the required major links listed above.

This current state of affairs leaves the entire theory of evolution at ground ZERO. In other words, if the long evolutionary journey proceeding from the first microbes to man were likened to sending a space probe on a prolonged complex mission into deep space, the trip could have never begun because the payload of non-living chemicals (dirt) has never been able to get off the launch pad without a chemical or biological vehicle!

As you continue it will be shown, in fact, that the 'technology' to launch this space ship is not available to naturalistic processes. The failure of this **absolute re-quirement** creates an impossible evolutionary beginning because dirt cannot possibly design, engineer and construct an organic mechanism equivalent to a "Saturn" rocket launch vehicle.

Kennedy Space Center Pre-Launch

Here's the mission:
We must send this payload on a
long journey into deep space.

???
but...but... we have no
launch vehicle!

What is the payload – anyway?

OH – that pile of dirt sitting
on the launch pad.

Illustration from composite of NASA images
Courtesy NIX

Unimagined Complexity

It was not too long ago when life at the cellular level was perceived to be somewhat simple. Just a few decades ago, a single celled organism was believed to be composed of "gelatin like" protoplasm confined within a membrane. The advent of the electron microscope combined with the development of many unique investigative research techniques has now revealed a cellular word of unimagined complexity. The resulting information explosion has created a growing demand for professional cytologists and microbiologists among the medical and pharmaceutical research labs. These scientists have demonstrated that a living cell is literally a universe within itself.

Each cell contains multitudes of subunits that in turn comprise complex systems, somewhat analogous to the organs and functioning systems of the human body. In addition, these components and systems are found to be essentially similar in all cells from bacteria to men. Another unexpected result of years of research indicates that even though cells have been identified with diverse functions, sizes and shapes, no evidence of any evolutionary sequence between them exists. [#2]

No scientist today would refer to **any** living cell as simple or primitive. In fact, no machine made by man can match the function and complexity of a self replicating cell.

LIVING CELL

Basic cell components
Courtesy: OSI Pharmaceuticals, Inc.

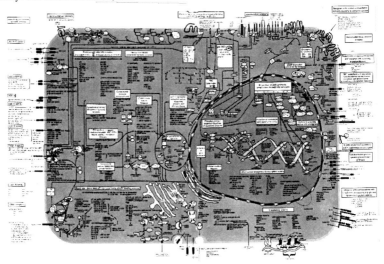

SATURN ROCKET

Saturn 1B cutaway &
Pre-Launch 1975

Courtesy: NASA
IMAGE
EXCHANGE

Which mechanism is more complex?

Dealing with these facts has created a mega problem for evolutionists, thus it has become necessary and popular to supply just-so stories on how these marvelously complex biologic machines could be accidentally assembled without the benefit of prior design, plan or purpose.

The best guesses only amount to speculation in hopes of finding a simple molecule, which is able to replicate itself. Keep in mind that any proposed self replicating molecule, of course, must function without the assistance of the amazing sustaining machinery that is only found within the manufacturing plant of a living cell! However, the ac-cumulated total of scientific investigation reveals that all life as we know it starts with and requires a fully functional complex cell factory.

If you choose to believe "In the Beginning...Dirt", ask yourself- **are you relying on facts or hope?**

> "...not one self replicating RNA (protein molecule) has emerged to date from quadrillions of artificially synthesized, random RNA sequences"[3]

1. Bergman, Jerry, 2000, Why Abiogenesis Is Impossible, *Creation Research Society Quarterly* 36(4):195-207

2. Denton, Michael,1986, *Evolution: A Theory In Crisis,* Adler & Adler, Bethesda, MD.

3. Dover, Gabby,1999, *Looping the Evolutionary Loop,* Nature:399

The Answer Key

TWO WAYS TO KNOW THE TRUTH

All Creation – one way to know the truth

Slow down a little and look around because all creation has been helping mankind to determine the truth. If you are uncertain of the correct choice, all you need to do is open yourself to the evidence that surrounds you every day. The scripture, from the book of Romans, directs us to look at the world around us:

> *From the creation of the world,*
>
> *God's invisible attributes, eternal power and Divine nature*
>
> *HAVE BEEN CLEARLY SEEN,*
>
> *being UNDERSTOOD through all that surrounds us,*
>
> *therefore, everyone is without excuse*   ***Romans 1:20***

With this in mind, I have provided a brief summary of some physical evidence which you have opportunity to observe every day or learn about for yourself. If Romans 1:20 is true, then *everyone* should be able to discern intelligent design and order throughout the earth and visible universe. In other words, you do not have to be a brain surgeon or trained scientist to discern creation. As an example, let us think through a few common situations. Have you ever walked through a creek bed or rocky desert foothills and looked down to pick up an object like this?

After examining it, you conclude this is not typical of one of the "naturally" formed, millions of stones, rock fragments or pebbles around you. This unique object is unquestionably an arrowhead. What is the significance of this find? Namely, the reason your senses and mind singled out this object from countless other rocks, which you have disregarded and trampled over, is because you classified this object as a **HUMAN ARTIFACT**. But wait a minute - Why? Obviously because your amazing brain can recognize and identify *Intelligent Design & Function.*
Oh, by the way, therefore...

> DESIGN AND FUNCTION REQUIRE A
> DESIGNER AND MAKER!

.

Similarly, you can discern that the existence of a computer **requires** a computer maker and I might add a programmer to make it functional. Corresponding to this example are all the living and non-living systems on earth, which *clearly* exhibit order, design, and function. These are the criteria which your brain uses to classify the origin of an object as a product of intelligent intervention or random natural environmental factors.

An Orrery: An early mechanical model of the solar system

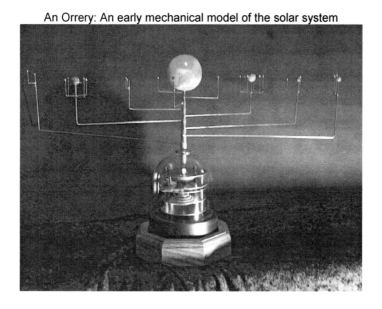

The story is told about a scientist, an atheistic friend of Sir Isaac Newton, who dropped in for a visit. Upon entering Sir Isaac's house, the friend noticed a mechanical model of the solar system. Having never seen such a marvelous device, which could duplicate the motion and position of the planets, he began cranking it. Observing the motion which mimicked the known planets of the solar system, he insisted on ascertaining "WHO MADE THIS?" Sir Isaac simply replied to repeated questioning - "no one," until his guest was totally frustrated. Finally, Sir Isaac responded in this manner; you stand here insisting that some one must have made this device because you clearly perceive intelligent design in this relatively simple model, however, you **REFUSE** to recognize the requirement for a designer for the corresponding, yet infinitely more complex, existing Solar System!

Question: Are you among one of those who somehow has missed this simple truth?

I believe one more example should be sufficient to help you understand that information, order, and function, *wherever observed*, require a designer. People who are oriented toward evolution often don't see evidence of purposeful design in living things, even though they do so for objects that are obviously man-made.

If I were to take a life-like artificial plant and have an evolutionist examine it, he would notice the imperfections and would invariably declare, "yes, it sure looks very authentic. It must have taken a great deal of skill and cunning to make this one - *but for certain, it's been manufactured*." There is no way this object could have formed by natural means. The odds of this happening are ZERO as far as this person is concerned. Now, isn't it amazing that a person can declare with certainty that the mock model had to be a product of intelligence but the infinitely more complex, self-reproducing plant, from which the model is only a crude copy, is **BELIEVED TO COME INTO EXISTENCE BY CHANCE**. Think now, if this were true, it must mean that any time we come in contact with something that is beyond man's capacity to make, then it must have come about by the undirected, accidental random process of evolution! But the odds of this happening must be much less than ZERO because of its obviously greater complexity. You know, if a person is capable of recognizing a particular artist solely by his style, how much more so the handiwork of a Creator?

LOOKING FURTHER . . .

ANY PERSON seeking the truth should carefully consider the sampling of evolutionary ideas and corresponding explanations presented in the next few pages. You may be surprised at how easy it is to understand the basic issues concerning origins.

The Truth Is Usually Simple

Theory A:

Since evolution has proceeded for millions of years (it is assumed), remains of intermediate biological forms should number in the millions. They should literally fill the museums and university collections. The perpetual change from simple to complex organisms should be obvious, if this theory is correct. Actually, the world should now be filled with untold numbers of "Star Wars" types of animals and sub-humans.

Observation:

After searching for over 100 years for the predicted intermediate transition fossils among the 200 million fossils identified to date, scientists have failed to produce any recognized conclusive examples - **NOT ONE!** This conspicuous lack of evidence is always referred to as missing links in the evolutionary literature. Even some evolutionary oriented scientists now suggest they will never be found because it is conceded that the vast known fossil record indicates they simply do not exist.

THE MISSING LINKS ARE STILL MISSING

But...but, what about all the transition fossils (links) which I saw for myself in the museums, science exhibits, school text books and T.V.? As a geologist, may I suggest that you take the time to examine these proposed links closely? Think- what did you really see?

Pictures, drawings, illustrations, mannequins and imaginative reconstructions

For example: I was touring the Arizona-Sonora Desert Museum when I came across an elaborate display which proposed to show the actual stages of a small dinosaur evolving into a bird. Multitudes of people viewed these elaborate transitions as evolutionary fact but **did they consider** that the entire exhibit was made from cast iron and not a single actual fossil! This scenario is typical of nearly all proposed transitions. Invariably, the display will lack a series of real fossils that clearly demonstrates any progression from one basic kind of animal to another, or animal to man.[#1]

THE FOSSIL LINKS ARE PROPOSED
AND BASED ON IMAGINATIVE THEORY

If you are not inclined to believe me, here is just a very small sample of what **evolutionary** scientists themselves have to say concerning the lack of confirmed transitions in the fossil record:

INVERTEBRATES CHANGING TO FISHES

"The geological record has so far provided no evidence as to the origin of the fishes..."   J.R. Norman, Department of Zoology, British Museum of Natural History [2]

FISH CHANGING TO AMPHIBIANS

"There are no intermediate forms between finned and limbed creatures in the fossil collections of the world" Gordon Rattray Taylor, former Chief Science Advisor, BBC Television [3]

REPTILES CHANGING TO BIRDS

"The origin of birds is largely a matter of deduction. There is no fossil evidence of the stages through which the remarkable change from reptile to bird was achieved." W.E. Swinton, British Museum of Natural History [4]

EVOLUTION OF PLANTS

"I still think that, to the unprejudiced, the fossil record of plants is in favor of special creation." Prof. E.J.H. Corner, Professor of Tropical Botany, Cambridge University [5]

LACK OF TRANSITION FOSSILS

"Despite the bright promise that paleontology provides a means of 'seeing' evolution, it has presented some nasty difficulties for evolutionists the most notorious of which is the presence of 'gaps' in the fossil record. Evolution requires intermediate forms between species and paleontology does not provide them." David B. Kitts, School of Geology and Geophysics, University of Oklahoma [6]

"...there are about 25 major living subdivisions (phyla) of the animal kingdom alone, all with gaps between them that are not bridged by known intermediates." Ayala and Valentine, [7]

"...the gradual morphological transitions between presumed ancestors and descendents, anticipated by most biologists, are missing." David E. Schindel, Curator of Invertebrate Fossils, Peabody Museum of Natural History [8]

Theory B:

Numerous fossils are believed to represent creatures which have become extinct millions of years ago. The precise dates for the origin and extinction of these organisms are often stated with certainty. These ancient extinctions are perceived as scientific data in support of the enormously long ages required of evolution, through the mechanism known as natural selection - "survival of the fittest."

Observation:

The fossil record clearly shows us that many more species existed in the past than are now living on planet Earth. However, extinction is a process of species reduction whereas evolution is proposed as a long-term process of diversification. Therefore we should now expect a greater variety of all living creatures as new life forms replace the vacant niches with more viable organisms (microbes to man). The problem becomes evident when even a casual examination of the fossil record documents a declining time-trend in the number of life forms rather than the expansive radiation predicted by evolution theory.

Extinction is not necessarily related to eons of evolutionary time. It is common knowledge that extinction can occur relatively quickly and is often caused by factors other than long term natural selection such as man's encroachment or by catastrophic changes within the environment. What is not commonly emphasized is that scientists have identified over **500 fossils** which have been claimed to be extinct, some for **as long as 400+ million years**, and yet identical organisms are found alive and well today.[#9] Such organisms are classified as "Living Fossils". There are many more of them around then previously acknowledged. Many fossils have been "tagged" with a specific genus and species biological classification which obviously should have been assigned to the equivalent modern (currently living) genus and species. Apparently, it has been common practice for the person naming the fossil to assign it a distinct name because of evolutionary oriented bias. Now that the living counterpart of these "extinct" organisms are being iden-tified, it is obvious that these organisms have not evolved over the imagined vast time spans of their existence into other, more complex creatures. Either evolution is a myth or these fossils are not really very old...**OR BOTH!** For example, paleontologists allege that particular extinct Graptolites are the index fossils that **positively** identify rock strata from a bio-stratigraphic zone of the Ordovician period. Index fossils are believed to be organisms whose evolutionary first appearance is precisely known and subsequent time span to extinction is very short. Therefore it is believed that they can be used as a definitive indicator of the age of the rock unit in which they are found. Graptolites are colonial marine animals consisting of one or more branches that look somewhat like ferns. Sedimentary rocks containing the index species of Graptolites are commonly cited in geology textbooks to be 300 million years old. Yet recently, these Graptolites have been discovered living in the south Pacific. Logically then, the age of Ordovician strata need not be hundreds of millions of years nor have Graptolites evolved into any other more complex organism over this proposed enormous span of geologic time.

A Graptolite Colony Fossil In Shale

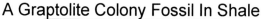

Photograph from U.S. National Museum

Likewise numerous other living fossils, some of which include time honored index fossils, are being discovered almost daily. A brachiopod also "extinct" is pictured below. Significantly, living fossils also include the alleged most ancient of all organisms. They are the single celled life forms that represent the candidates of the earliest evolutionary stage emerging from a chemical soup. These fossils have been identified in rocks of proposed ages ranging from 3.4 to 1.5 billion years old. The assigned dates for these micro-fossils initially may have seemed logical but now prominent scientists have concluded that these fossils are essentially identical to living bacteria and blue-green algae. Why then would these organisms, having undergone billions (perhaps trillions) of reproductive cycles, remain unchanged if evolution indeed has been active for over a **billion** years? This consideration, along with numerous other types of living fossils, relegate the idea of continuous evolutionary organic progression to the category of an imaginative story.

Space only allows a few examples of living fossils below but numerous plants and animals are documented which include amphibians (frogs), reptiles (turtles, crocodiles), and many mammals which are basically identical to corresponding ancient extinct fossils.

The recent publication of the book "Living Fossils"[#10] actually represents a formal scientific falsification to the concept of biologic evolution. This extensive work of Dr Carl Werner clearly documents the lack of evolutionary change in **hundreds** of life forms over the assumed vast period in excess of 100 million years. His extensive field work and research of a comparison of numerous fossils from the Triassic, Jurassic and Cretaceous geologic Periods, known as the Age of the Dinosaurs, confirms that no significant biologic change has occurred in nearly every animal Phyla or plant Divisions. The Genera and often even the species of the dinosaur era organisms have remained essentially unchanged and therefore numerous fossil plants and organisms can be clearly identified as equivalent to modern life forms.

Most paleontologists now readily admit that the fossil record is characterized by abrupt appearance of life forms with no trace of evolutionary ancestors. These organisms have remained essentially unchanged throughout the geologic strata wherever they are found. This condition is well known in the scientific literature as "Stasis".

Living fossils; therefore, represent "hard" evidence of the **stability of life forms throughout the geologic history of the Earth.** Furthermore, the absence of these fossils in geologic formations, interpreted as extinctions, does not necessarily represent evidence for vast intervals of time. This, of course, is unwelcome news to evolutionary theorists, particularly when the discovery of a living fossil has been previously used as an index to "support" the evolutionary time scale. Now you can discern why the disclosing of numerous "living fossils" is not emphasized in the popular scientific magazines, T.V. programs, schools or the news media and yes, even the majority of museums. --OH!

> THE REALITY OF LIVING FOSSILS
> AND
> THE THEORY OF EVOLUTION DON'T MIX

LIVING FOSSILS

Proposed Extinct Species Ranging From Hundreds (Top)
To Tens of Millions of Years Old (Bottom)

Living Organism   *Fossil*

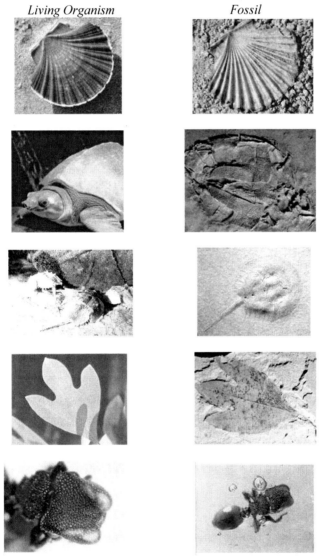

Courtesy Dr. Joachim Scheven
Labendige Vorwelt Geological Museum
Grunstadt, Germany

50

Theory C:

A popular evolutionary notion had maintained that the human body contains as many as 200 useless vestigial organs left over from our evolutionary development. (Ear, muscles, tonsils, appendix, etc.) As a result of this belief, these organs have been routinely surgically removed for decades and in some cases people have died needlessly because of this.

Observation:

Almost all of the so-called vestigial organs are now *known to perform essential functions.*[#11] Doctors are now reluctant to remove even tonsils. This shows a definite shift in thinking from the notion of random development to **purpose and function** for these organs.[#12] Logically, therefore, the human body points toward creation. In addition, don't hesitate to ask your anatomy teacher "Where are the nascent organs?"

(emerging organs in the evolutionary process of forming)

According to evolution, there should exist many organs which are not functional yet because they are still in the process of evolving, being only partially developed. Hmmm - that's right, there are none in the human body nor has any other such vestigial or nascent organ been identified in any organism. You mean there are no nascent organs in dogs, beavers, snails or elephants? Oops - another fact we should be aware of …

ALL LIVING ORGANISMS APPEAR TO BE FULLY FORMED AND FUNCTIONAL

Conclusion: If life did evolve, it would indeed be reasonable to expect nascent and vestigial organs in mice, snakes, cats, fish… and man.

but - REALITY REFUTES THEORY!

Theory D:

The ability of germs to "develop" immunity to antibiotics and insects to develop resistance to insecticides are often claimed as classic examples of evolution. Children's popular books and school text books commonly present these examples as *facts* of evolution.

Observation:

The science of genetics has clearly shown that resistant bacteria or insects do not result from an increase in the organism's complexity or genetic information. The latter is required to verify evolution. In other words, no new superior or complex organism has evolved. Increased resistance or survival results from an increase in reproduction of **already existing,** more resistant varieties of the **SAME** organism. For example, the more resistant variety will simply proliferate as the common less resistant 'germs' are eliminated by an antibiotic. At the end of the day, the basic form of the remaining germs is still the same species and the same germs. The surviving resistant insects are still the same sort of insect. The bacteria, for instance, did not turn into, or evolve into, an immune beetle. The plant eating worms, which were sprayed with insecticide, did not turn into resistant roaches!#13 &14 All knowledgeable scientists today would not use this as an example of evolution, because the science of Genetics says NO!

In fact it is well known that the more resistant varieties will quickly die off and the common organism will proliferate again when the antibiotic or insecticide is removed because they are not the evolutionary product of a more superior strain but rather are of lower viability (inferior) and unable to survive under normal environmental conditions.

INCREASED RESISTANCE IS NOT EVOLUTION

Theory E:

Evolutionists say they "know for sure, the Earth is 4.5 Billion years old. Given this much time, anything could have happened, and it did." It is important to under-stand that In order for the theory of evolution to be even remotely plausible, it must demonstrate that hundreds of millions of years of Earth's history was available.

Observation:

First of all, *everyone* should be aware that there are **NO - NONE- ZERO** scientific laws, principals or demonstrated valid scientific experiments which provide a mechanism for life to construct itself from nonliving chemicals - **No matter how much time is available!**

Current research using high tech instruments now assures us that that there is no such thing as the simple cell. The analogy between raw chemicals assembling themselves into a reproducing cell can be comprehended as being similar to nuts, bolts and rivets eventually assembling themselves into the Space Shuttle if placed in a tumbler.

> Only a great leap of faith could erect the imaginary bridge which spans the gap between molecules and life

Second, we do not know that the Earth is 4.5 Billion years old. This estimate is based upon theoretical geochemical analyses of rocks and meteorites. In fact, the preponderance of evidence accumulated from numerous natural processes (75+) has been published by scientists that point to a very young Earth. This data are clearly more aligned with the biblical chronology of Earth history. The reason why most people are unaware of these is because the media rarely refers to the many scientific estimates that yield an age of less than 10,000 years. But you would have to ignore all of them, if you choose to believe in an ancient Earth. In addition, the evolutionary community simply can not acknowledge this research because it obviously leaves no room for the required millions of years to accommodate biologic evolution. Remove this "icon" of evolution of deep time and the theory fades to just a myth.

Stacked against a multitude of young chronometers are the **two** primary methods used by the evolutionary com-munity to support the idea of an old Earth.

#1 - The **ASSUMPTION** that the Earth has been around for billions of years. The support for this assumption is the principle of Uniformity. I'm sure you heard of it stated as "the present is the key to the past". This principle was proposed in 1830 by Charles Lyell, an English attorney, when he published his ideas regarding the geologic history of the Earth.

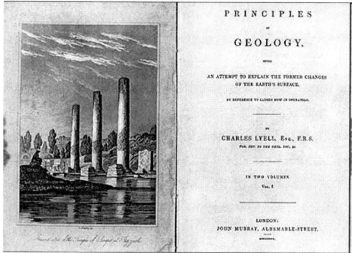

As a result, the principles which he proposed became very popular, eventually convincing the geological establishment to abandon the long held belief in the biblical flood. As a result uniformitarianism has been accepted as the under-pinning of all geological thinking until relatively recent. Starting with Geology 101 at any university, students are taught that the character and rate of geologic processes which we historically have measured and observed have been operating essentially unchanged throughout all of Earth's history. As a geology student myself, this idea sounded reasonable to me. After all, today we observe that rivers slowly erode land surfaces and gradually transport the sand downstream to the ocean where they build deltas. The sediments also are eventually spread gradually out on the sea floor, where a variety of bottom-dwelling creatures could perhaps be occasionally buried and then fossilized. By the way, scientists freely admit that there is nowhere on this planet, whether under the oceans, lakes or on land, that extensive fossil beds are forming today!

So then, how has the age of the Earth been traditionally been estimated? Generally, the *current* slow rate of accumulation of rock sediments observed *today* is multiplied by the average thickness of the geologic column (rocks comprising the Earth's crust). It is reasoned that this yields the estimated eons of time required to form the rocks of the Earth's crust. Use of this method is like seriously trying to convince someone that you are going to be reading your Bible *continuously* for the next 100 years. Your proof is showing them a camera snapshot made of yourself at your desk reading a Bible! The **current** photo, supposedly, represents proof of your activities for the next 100 years. If you believe the logic of this scenario, then you are also at liberty to accept the **assumption** that all the geologic processes which are observed today (a snapshot in time) have always transpired at the current relatively slow rate.

The fallacy of the uniformitarian principle is becoming more evident, however, as the physical evidence being published by credentialed young earth geologists can no longer be ignored. It is obvious from detailed studies of the earth's crustal rocks that this planet has experienced numerous catastrophic episodes of sedimentation, erosion and volcanism of much greater magnitude than the current conditions exhibit. Many geologists now recognize that this traditional idea - 'the present is the key to the past' – is no longer valid. Today's observations, like a snapshot in time, cannot be extended back indefinitely to interpret the history and age of the Earth. Geologists can discern from carefully deciphering the rock formations of the Earth that this planet has definitely experienced rapid catastrophic world wide geologic events which are **unprecedented** today.

For example; the rock record reveals evidence of the pre-existence of:

➢ Enormous volcanoes that deposited huge volumes of ash over millions of square miles.

➢ Extensive sandstone and conglomerate rock formations which have been deposited over continental areas of hundreds of thousands of square miles. Current sedimentation processes can not account for these deposits.

> Mega earthquakes have moved entire mountain ranges and have apparently destabilized vast ancient surface lakes. Enormous volumes of water released from these lakes resulted in rapid erosion which formed some of the largest and deepest canyon systems on Earth.

> More recently, it has been concluded by geologists that particular sedimentary deposits which have previously been difficult to explain are now interpreted as having been formed by highly energetic ancient storms. These hypercanes pro-duced winds estimated at 500 miles per hour and may have lasted for weeks. This is opposed to current hurricanes which last a matter of days with winds ranging from 75 to about 150 miles per hour.

> It is apparent that the intensity and rapidity of former geologic events has **subsided** to the current relatively quiescent rates. Occasionally this planet does experience what we would consider a catastrophic event such as a volcanic eruption, but these are clearly not even close to the magnitude revealed in the geologic record.

It is therefore becoming more obvious to many geologists that much of the structure of the Earth's rock strata and surface topography is not necessarily the result of slow gradual sedimentation over millions of years of geologic time. Obviously, then the age of the Earth based on current processes has been greatly overestimated by the simplifying theorem "the present is the key to the past".

#2 - The **ASSUMPTION** that some radioactive isotopes can be used like precision clocks to provide absolute ages for rocks in which they are found.

Creation scientists recognize that the theory of radiometric dating is basically theoretically correct for dating imaginary ideal samples on paper. As a graduate geology student at Syracuse University, the author was quite proficient at calculating these dates because I had a strong background in mathematics as a former mechanical engineer. What I did not realize, however, is that the dates which I calculated were only theoretically correct but did not represent reality. After pondering all the evidence which supports a young Earth of thousands of years, not billions, I have since realized that there exists a vast difference between a text book problem and a rock sample exhumed from the earth's dynamic crust.

Radioactive Decay

(parent isotope)   *(daughter isotope)*

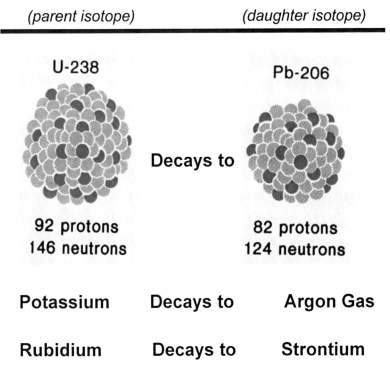

U-238   Decays to   Pb-206

92 protons   82 protons
146 neutrons   124 neutrons

Potassium	Decays to	Argon Gas
Rubidium	Decays to	Strontium

I believe the following explanation will make it easy for the reader to understand why radioisotope dating is unreliable. The problem is that a text book rock sample represented as a closed system and isolated from the surrounding environment by drawing a box around it on paper is not representative of the real world! The dating of an ideal text book sample is accomplished by solving mathematical equations relating to assumed constant decay rates (half life) and measured ratios of initial and end products of the radioactive decay process.

The mathematical expression that relates radioactive decay to geologic time is:

$$t = \frac{1}{\lambda} \ln\left(1 + \frac{D}{P}\right)$$

> *t* *is the age of the rock or mineral sample*
> *D* is number of atoms of the daughter products in the sample,
> *P* is number of atoms of the parent isotope in the sample
> *λ* is the decay constant of the parent isotope; related to the half live of the parent isotope
> ln is the natural logarithm to base e

The exercise requires that the chemical system enclosed in the "box" has been an ideal closed system throughout its entire existence. (refer to the illustration) This is relatively straight forward on paper. Just draw a box around the sample. This ideal paper model provides the constraints to chemically isolate the sample from the rest of the world and allow someone to calculate the theoretical age of the isotopes inside the box.

Mathematical modeling is a common technique used in the fields of thermodynamics, structural mechanics, physics, etc... Many ideal models have proven to be very practical and yield useful results for scientists and engineers. Even though they do not always equate to experimental test results exactly, they may be successfully used through the use of multiplication coefficients (fudge factors) or calibration curves to accurately mimic real world conditions.

Comparative illustrations of the theoretical closed system and the open system of Earth's rock strata are provided on the following pages.

IDEAL RADIOISOTOPE CLOSED SYSTEM MODEL

Rock sample inside safe

Door has never been opened

Isolated from surrounding environment until sample is removed for testing

<u>ASSUMPTIONS REQUIRED TO DATE SAMPLE</u>

1. Constant radioactive decay rate
2. No gain or loss of original parent isotopes (door is locked)
3. No gain or loss of decay products (nothing can penetrate the concrete lined safe)
4. Quantity of similar decay products present at start assumed to be zero.

DYNAMIC OPEN SYSTEM OF THE EARTH'S CRUST

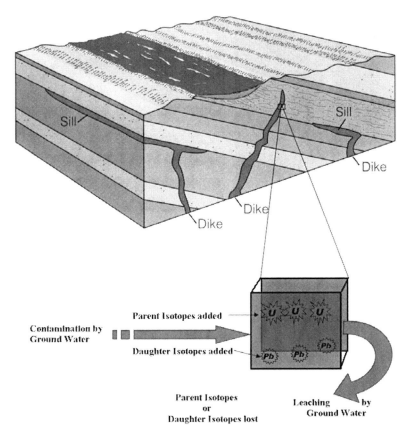

Rock Sample Taken From Basalt Dike

Sample is usually subject to many factors:

- ➤ Chemical enrichment or leaching of parent element
- ➤ Loss or gain of decay products by leaching or diffusion
- ➤ Net gain or losses unknown
- ➤ Rarely conforms to Ideal Model

Question: Is the Ideal model at least close enough to reality permit estimating accurate or consistent ages?

Answer: If you check out the numerous articles on the internet regarding radioisotope dating you will find that many scientists endorse the method as reliable. They will invariably point out that additional techniques are now in use to mathematically account for the indefinite open system history of the rock or mineral being dated. Although improvements have been made to the basic ideal model, the fact remains that the unknown factors consistently result in **grossly conflicting** age determinations.

This is, in part, why a growing number of scientists and technical journals now report that the various radioactive clocks are plagued with severe problems when applied to **actual field samples**. In other words, there is a great difference between theory and the useful practical application of dating real rock samples which have been subject to thermal and mechanical mixing during the molten magma state and chemical weathering after solidifying. This problem becomes obvious when the dates obtained are *nearly always inconsistent* after the various isotope dating methods on the same sample are compared with each other.

Isn't it logical to expect several clocks, all in good working order, to "tell" the same time? Invariably, a rock sample dated by three independent radiometric clocks will yield **THREE** completely different dates. Which one is correct - if any? Adding to this confusion is the generally unspoken fact that many test samples have been dated, whose ages are already **KNOWN** because they were formed during recent or historical volcanic eruptions. When samples of these rocks are dated to test the **validity** of these radiometric clocks you would expect, of course, the results to coincide with the known age of the sample - but they do not. The calculated dates are in-variably totally erroneous. Well, how much in error; - hundreds of years or perhaps a few thousand years? Would you believe, tens to hundreds of millions of years in error! Would you want such a watch for your birthday?

Objective analysis has shown *repeatedly* that:

THE RADIOMETRIC CLOCKS SIMPLY DO NOT PROVIDE ABSOLUTE DATES BECAUSE THE VARIOUS RADIOISOTOPE MODELS CAN NOT REPRESENT OR PROVIDE AN ACCURATE TIME-HISTORY OF ROCK SAMPLES OF KNOWN AGE

This conclusion is not drawn from speculation but from actually observing THE CONFUSION WHICH RESULTS FROM TRYING TO SORT OUT THE NUMEROUS INCONSISTANCIES OF THE VARIOUS ISOTOPE DATING METHODS.

It's simple - you see a man standing on the street, selling watches. He is trying his best to convince you these are superb Rolex time pieces, in working order, that he personally synchronized at the same time last week. Ok, you say, "show me six of them." You **OBSERVE** that every one of them indicates a vastly different time. Would you "buy" his story???

THE TRUTH IS OBVIOUS ...

and you decide to keep your own Timex !!!

Radioisotope Evidence Supporting a Young Earth

Recognizing multiple weaknesses in the traditional procedures used in radioisotope dating, a group of scientists formed a team in 1997 to investigate these problems. The research project is referred as RATE. (Radioisotopes And The Age Of The Earth)

Their research spanned about eight years. All financial support for the project was supplied by donations from private individuals and creation research organizations. Comprehensive documentation of this project is available from two sources:

"Radioisotopes And The Age Of The Earth" (Vardiman et al., 2000) and;

"Results Of a Young Earth Creation Research Initiative" (Vardiman et al., 2005)

In lieu of these technical publications, a readily understandable popular summary of this extensive investigation has been published by one of the RATE scientists, Dr. Don DeYoung. "Thousands Not Billions".[15] Several significant conclusions from this research project are summarized below:

1. There has been a widely held misconception for several decades that Carbon-14 dating does not support the young Earth view. However, the RATE team has documented many instances of the presence of C-14 in significant quantities in ancient fossils, coal and diamonds. These findings were **not expected** and prior reports of such findings have previously been essentially ignored by the evolutionary community. The half life of the radioactive element C-14 is relatively short, being only 5,720 years. Any sample whose age is believed to be greater than about 50,000 years should have no detectable C-14 remaining. Therefore the presence of C-14 in specimens which are traditionally considered to be millions to hundreds of millions of years old provides powerful evidence that clearly supports a very limited age of the Earth.

This data was obtained by performing state of the art tests on various datable carbonized fossils which were collected from strata throughout the traditional time range of the Geologic Column. The presence of C-14 is in **direct conflict** with the uniformitarian assumption of slow accumulation of the sedimentary rocks over eons of "deep Time".

2. It has been duly noted that most results determined from the isotopic dating of rocks have persistently indicated an apparently ancient age for the Geologic Column of strata around the Earth. Although, as previously revealed, the results are inconsistent and have been shown to be incorrect for samples of known age, this has remained a definite challenge to the contention of a young Earth by creation scientists. Young Earth scientists have lacked an adequate explanation for the presence of a relatively large ratio of daughter products to parent isotopes found in most samples. This would appear to confirm a large amount of radioactive decay which, in turn, requires a long span of geologic time.

This apparent anomaly revolves around the assumption of constant radioactive decay rates. For the most part, nearly all scientists have presumed that the rate of radioactive decay has remained constant throughout the history of the Earth. It should be noted that this critical assumption is only based on the short span of current observations. Remember the mathematics of isotopic dating models are based on the assumption that nuclear half-lives remain stable and unchangeable.

The RATE researchers have made some pertinent discoveries through the precise measurement of just how quickly Helium atoms flow through Zircon crystals and the surrounding minerals which enclose these crystals in granite rock. The Helium contained within the Zircons accumulates as a product of radioactive decay of Uranium. Uranium atoms are similar to Zirconium atoms and therefore may be incorporated in the crystal structure of the Zircons.

Reactions within a Zircon Crystal

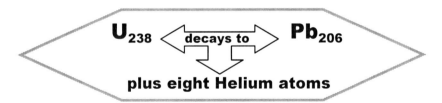

The Zircons were removed from granite cores located in deep test borings about 2.5 miles below the surface of New Mexico. Standard radioisotope dating indicated that the granite is about 1.5 billion years old. This apparent age was confirmed by the RATE investigators, thus indicating that a substantial amount of Uranium had decayed to lead. However, accurate measurements of the relatively **large quantity** of Helium gas remaining in the Zircon crystals required that the crystals be no older than 6,000 ± 2,000 years. This date could be determined with precision because the known rate of diffusion of the very small Helium atom through Zircon crystals had been accurately measured by the RATE team. If the crystal was, in fact, 1.5 billion years old very few Helium atoms would remain. So the RATE scientists are faced with two opposing dates for the granite sample. On the one hand, a large amount of Uranium had decayed to Lead, indicating the passage of deep time. However, the presence of excess Helium indicates a young age.

A reasonable explanation for these inconsistent findings is that the data implies that the long accepted assumption of an enduring constant isotope decay rate may no longer be valid. It is evident that the unexpected excess Helium must have been the product of accelerated Uranium decay at some time during a period of 6000 years. This compelling evidence, again, can be added to the long list of collaborating evidence pointing to a young Earth.

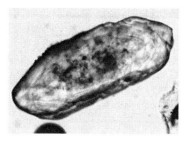

Micro- photo of Zircon crystal: about the size of a small sand grain.
(Photo from Wikipedia.org/wiki/Zircon)

There is no lack of information available to **whosoever** desires to compare the credibility of evolution with creation. I believe the examples given are sufficient to lay a solid foundation upon which you can reach a rational conclusion or, at least, encourage you to take a closer, more critical look at the issues of origins.

The later sections of this book will provide further details concerning the current state of knowledge related to life's origins. In addition, a list of related web sites and interesting publications are included to encourage further self investigation.

REFERENCES

1. Gish, Duane, T., 1996, *Evolution: The Fossils Still Say No,* Institute for Creation Research, El Cajon, CA

2. Norman, J.R., 1975, *Classification and Pedigrees: Fossils in History of Fishes.* Greenwood, P., Editor, 3rd edition, British Museum of Natural History, London, p.343

3. Taylor, G.R., 1983, *The Great Evolution Mystery,* New York, Harper & Row, p.48

4. Swinton, W.E., 1960, British Museum of Natural History, *The Origin of Birds,* Chapter 1, in Biology & Comparative Physiology of Birds, Marshall, A.J., Editor, Vol. 1, Academic Press, New York, p.1

5. Corner, E.J.H., 1961, *Evolution in Contemporary Botanical Thought,* Macleod & Cobley, editors, Botanical Society of Edinburgh, p.97

6. Kitts, David, B., 1974, *Paleontology and Evolutionary Theory,* Evolution, Vol. 8, (September), p.467

7. Ayala, F., and Valentine, J., 1979, *The Theory and Processes of Organic Evolution,* Benjamin Cummings Pub. Co., Menlo Park, CA., p.258

8. Schindel, David, E., 1982, *The Gaps in the Fossil Record,* Nature, Vol. 297(May), p.282

9 Scheven, Joachim, *Living Fossils: Confirmation of Creation,* Creation Videos, (VHS) Available thru AIG web site listed at end of book

10. Werner, Carl, 2008, Living Fossils, New Leaf Press, Green Forest, Ar

11. Bergman, Jerry and Howe, George, 1990, *CRS Monograph Series 4,* Creation Research Society, St. Joseph, MO

12. Maas, Frank,1994, *Immune Functions of the Vermiform Appendix,* Third International Conference On Creationism, Ed., Robert Walsh, Creation Science Fellowship, Inc., Pittsburg, PA, p.335

13. Wieland, Carl, 1997, *Superbugs,* Creation ex nihilo, Vol. 20(1); pp.10-13

14. Penrose, Eric, 1998, *Bacterial Resistance to Antibiotics,* Creation Research Society Quarterly 2 (35); pp.76-83

15. DeYoung, Don, 2005, *Thousands Not Billions,* Master Books, Green Forest, Ar

ANOTHER WAY TO KNOW THE TRUTH

Revelation

I bet you have heard something like this before, "No one was there at the beginning to observe the origin of life, that's for sure"! Or then, again - are we so sure?

Having reviewed some of the physical evidence, let's look at the second way in which every individual can know the truth. It has been previously pointed out that the laws of physics require an origin for the universe. In addition, everything that we observe in the living world points to an intelligent designer.

So let us look into another line of reasoning. Mankind is fortunate in that we have been blessed with direct revelation from the *only living witness* to creation "from the very beginning". God, the prime designer - has Himself given to us His own eyewitness account. This fascinating insight is recorded in the first chapter of Genesis. The fact that the entire Bible opens with the revelation of creation, unquestionably speaks of its major importance as foundational to understanding the plan, purpose and origin of life.

Understand that the creation revelation is not the product of any man, as no man was there. It originates from **THE MIND OF GOD**, is received from the **WORD OF GOD** (the Bible) and indeed it corresponds to all that we can examine in the physical world around us.

74

Think - where does science get its information from, anyway? A body of knowledge known as science is really only natural revelation. And doesn't natural revelation consisting of scientific facts, data and experiments come from observing the handiwork of God all over the Earth? Therefore, let me ask you: If God is the creator of the universe, would you expect natural revelation (science) to contradict biblical revelation?

Would you expect the Hand of God to conflict with the mind of God?

How, then, can any speculations which you have read or will hear in the future regarding so-called scientific evidence, which contradicts God's Word, represent true science? The **facts of science** have always been found to agree with and compliment God's account of creation, origins, and man's history. I believe you will see this more clearly as we now look at the awesome revelation of creation given to us as a gift in the first three opening verses of the Bible.

Scripture starts with the origin of the visible universe and its chief tenant - man. Masses of people, including scholars, philosophers, scientists, and theologians have pondered the intriguing creation account for millennia. All attempts to determine the origin and purpose of life apart from the Creator's revelation are limited to human reason. Philosophy can never produce absolute answers. The best which human logic can produce is a multitude of subjective opinions followed by more questions. Unlike the bizarre fantasies and religious myths of non-Christian cultures, the Bible is unique in its straight forward presentation. Multitudes have accepted it as truth, including many of the greatest scientists of recorded history. Apparently, all the persons listed below were Bible believing scientists who endorsed the Bible's account of creation as completely compatible with scientific observations. Only a partial list is included here, but you should take note that many of these men were responsible for the key breakthroughs of modern science and are considered the "founding fathers" of the major scientific disciplines, which are recognized today.

It should be clear from this table that...

Statements to the effect that creationists are not
"real or credible" scientists are simply
NONSENSE

Key Contributions of Bible Believing Scientists*

Scientist	Discipline founded or key discovery / invention
Leonardo da Vinci (1452-1519)	Experimental science; Physics
Francis Bacon (1561-1626)	Scientific method
Johann Kepler (1571-1630)	Scientific astronomy
William Petty (1623-1687)	Statistics; Scientific economics
Blaise Pascal (1623-1662)	Hydrostatics; Barometer
Robert Boyle (1627-1691)	Chemistry; Gas dynamics
John Ray (1627-1705)	Natural history
Nicolas Steno (1631-1686)	Stratigraphy
Isaac Newton (1642-1727)	Dynamics; Calculus; Gravitation Law; Reflecting telescope
William Derham (1657-1735)	Ecology
John Woodward (1665-1728)	Paleontology
Carolus Linneaus (1707-1778)	Taxonomy; Biological classification system
Richard Kirwan (1733-1812)	Mineralogy
William Herschel (1738-1822)	Galactic astronomy; Uranus
John Dalton (1766-1844)	Atomic theory; Gas law
Georges Cuvier (1769-1832)	Comparative anatomy
Humphrey Davy (1778-1829)	Thermokinetics; Safety lamp
John Kidd, M.D. (1775-1851)	Chemical synthetics
David Brewster (1781-1868)	Optical mineralogy; Kaleidoscope
William Prout (1785-1850)	Food chemistry
Michael Faraday (1791-1867)	Electro magnetics; Field theory; Generator
Charles Babbage (1792-1871)	Operations research; Computer science; Opthalmoscope
Samuel F.B.Morse (1791-1872)	Telegraph
William Whewell (1794-1866)	Anemometer
Joseph Henry (1797-1878)	Electric motor; Galvanometer
Matthew Maury (1806-1873)	Oceanography; Hydrography
Louis Agassiz (1807-1873)	Glaciology; Ichthyology
James Simpson (1811-1870)	Gynecology; Anesthesiology
James Joule (1818-1889)	Thermodynamics
George Stokes (1819-1903)	Fluid Mechanics
Rudolph Virchow (1821-1902)	Pathology
Louis Pasteur (1822-1895)	Bacteriology; Biochemistry; Sterilization; Immunization
Gregor Mendel (1822-1884)	Genetics
Henri Fabre (1823-1915)	Entomology of living insects
Lord Kelvin (1824-1907)	Atlantic cable
William Huggins (1824-1910)	Astral spectrometry
Bernhard Riemann (1826-1866)	Non-Euclidean geometrics
Joseph Lister (1827-1912)	Antiseptic surgery

Scientist		Discipline founded or key discovery / invention
Joseph Clerk Maxwell	(1831-1879)	Electrodynamics; Statistical thermodynamics
P.G. Tait	(1831-1901)	Vector analysis
John Strutt,		Similitude; Model Analysis;
Lord Rayleigh	(1842-1919)	Inert Gases
John Ambrose Fleming	(1849-1945)	Electronics; Electron tube; Thermionic valve
William Ramsay	(1852-1916)	Isotopic chemistry, Element transmutation

*Adapted by permission:
Henry M. Morris, *Men of Science - Men of God*, Green Forest, AR: Master Books, 1988, p.99-101

Does the implication of this list surprise you? Probably so; because this information is not something brought to light by the world of public education. Secular textbooks conveniently do not point out that these prominent scientists were Bible believing Christians who accepted the creation account as written. Furthermore, there exists in the United States alone an estimated 10,000+ scientists who believe the facts support the biblical creation account. Their academic credentials are either at the masters or doctoral level and they span every conceivable scientific discipline. Many are members of growing professional creation science organizations and work in the cutting edges of technology.

Well, does it really matter what you believe about heaven, earth and the origin of life? Why not just let Jesus Christ answer that question:

> *"If you do not believe the writings of Moses*
> *HOW WILL YOU BELIEVE MY WORDS"*
> *John 5:47*

This concise statement is actually an intense spotlight that can illuminate the path of any inquisitive person to the true source of origins. It is fundamental to man's quest for understanding because the basis for **all** Christian doctrine regarding each person's need for the Savior originates in the opening chapters of Genesis... written by Moses.   For without God's marvelous revelation of His desire to create man in His own image with a free will and the subsequent decision of man to disregard God's word, there would be no conceivable basis for understanding the Gospel of the Cross or the grand purpose of the coming Messiah – Jesus Christ. If man simply evolved from a lifeless pre-biotic chemical soup - and that by chance - it follows then, in the mind of men, that life would have no ultimate purpose and the Gospel that is preached would be irrelevant. Think about it, no one, not even the deep thinkers or philosophers, can determine the purpose of life unless the revelation that God created man to enjoy an eternal relationship with Him is understood and received.

No man, whether a scientist, philosopher, guru, or minister can provide you with the ultimate knowledge of origins because **NO MAN WAS THERE** !

But, what about the Scientific Method? Surely science can provide this knowledge.

NOT SO!

Unfortunately, the man on the street doesn't realize that the Scientific Method is severely limited to proving only those things that are:

➢ *Repeatable*
➢ *Observable*
➢ *Testable*

It is obvious that the initial singular origin of matter or life was not observed nor can it be repeated, measured or tested. Therefore, you should be aware that **no scientific theory exists** nor is any secular attempt made to explain the origin of matter. The only alternative for man is to believe, by faith, in eternal existence of matter or the creation account given by an Eternal God.

A person should have no concern about accepting the Genesis account of creation as absolute TRUTH because it does indeed conform to the established laws of science. Multitudes of scientists, present and past, have recognized this correspondence.

With this in mind, there is no need to be influenced by the ever changing man-made theories which appear to conflict with the eternal Word of God. Numerous falsely conceived theories concerning origins have already come and gone but the Word of God remains creditable. The good news is that **technology is actually verifying the creation account and the world- wide flood with each advance in modern scientific investigations**. For example, the world of microbiology has expanded with the unveiling of the enormous complexity of a single living cell. No evolutionary explanation is remotely possible for explaining the origin of DNA. DNA is now understood to be the ultimate computer program that directs the trillions of building blocks of life to form a living organism.

In addition, recent laboratory and field studies in Geology are unveiling a picture of rapid worldwide deposition of the fossil bearing sedimentary rocks, corresponding to the biblical account of the world wide flood. Anyone open to the truth will have to admit that the evidence is greatly in favor of a catastrophic universal flood, rather than slow accumulation of rocks for millions of years. But this growing body of evidence should be no surprise, for scripture says that:

"Darkness will never overtake light" **John 1:5**

The truth of God's word will always be evident, no matter how much deception is cloaked as fact.
Remember this scripture?

"Where sin increased, Grace abounded all the more."
Romans 5:20 (NAS)

Yes, though sin is shown to be wide and deep, thank God His grace is wider and deeper still! **Romans 5:20** (Phl)

Although the Great Deceiver has manipulated multitudes into believing the lie of evolution and long ages of the Earth, by the Grace of God, modern studies in every area of research are refuting this darkness and bringing light to **whosoever** is open to the truth. Let's look at the opening verses of the Bible - God's creation revelation.

The Creation Account

Note that it is not a coincidence that the inspired Word **STARTS** with the revelation of creation. As pointed out previously, the prime importance of understanding the truth of creation is certainly emphasized by its significant first place in the Bible. Thereafter, the Word of God repeatedly pronounces God as Creator of the heavens and earth.

Now, check this out! Right from the opening of scripture, **God makes no attempt to prove that He exists**. Genesis opens by taking this fact for granted!

"In the beginning - GOD -"

The creation revelation starts as though the existence of God should be so obviously evident that "ONLY A FOOL WOULD SAY THERE IS NO GOD!" **(Psalms 14)**

VS:1 "In the beginning..." If some event has a beginning or an end, it must be related to a time frame of some sort. Well, I believe that you are simply being told here that TIME, as we understand it on earth, **starts here**. God is even the author of time. This time frame exists only in the physical universe and can be measured. Before the creation event, a watch would have had no relevance. The event revealed to you in VS:1 is the point in eternity where watches could start ticking off TIME. In other words, time had a beginning and apparently time, as we measure it, will have an end (1 Peter 4:7 & 2 Peter 3) when we enter a new phase of eternity with the Lord. In addition, we can conclude from the first verse that, because God already existed "In the beginning," therefore, God precedes time and, of course, we can understand from the Word of God that He is not time dependent but transcends time - dwelling in eternity.

Now if time had a starting point, as indicated in the very first verse, does this agree with science? Yes, indeed! Even the general theory of relativity, proposed by modern physics, indicates that time had a starting point. The origin of time is represented by "T_o" in the mathematics of physics.

Concerning the first occurrence of the Divine Name "God" in scripture, the Hebrew *"Elohim"* relates to His creative omnipotence. The *"im"* is actually the Hebrew plural but it is rendered singular when referring to God Himself. Could this be an initial suggestion to us of the plurality of the God-Head...His triune nature?

*"In the beginning God **created...**"*

The Hebrew "bara" is translated "created". Its intent conveys the act of God whereby He brings into existence that which had no previous existence. You may have heard of the Latin expression -

Creation - Ex Nihilo
meaning
Creation - From Nothing

Now, think about it. Does God really create from nothing? Yes and No. Indeed the Bible reveals that God did not require or use any physical material to create the universe. But He did use something, His creative Word.

That's why we are told in Hebrews 11:3 (NAS)

By faith we understand ... that the worlds were

prepared by the word of God, so that

what is seen was not made out of things which are visible.

(i.e. physical matter)

So what you have here, in the opening verses of the Bible, is the fundamental revelation of the creative power of the Word of God.

Let's consider the word Heaven(s). In VS:1, "In the beginning God created the heavens and the earth (NAS)... the original Hebrew word for heavens is *Shamayim*. (Like *Elohim*, it could be translated in the singular or plural) What you need to understand is that the word heaven, as used here, does not refer to the stars or celestial bodies at this time in the initial verse of the creation account. You can see that the sun, moon and stars were not yet created until day four (VS:6). The word heaven(s), in verse one, simply corresponds to our modern word for "SPACE."

It is interesting to notice that when Moses summarized the inspired creation account in Chapter 2:1, you can see that the heavens do not include, and it is distinguished from, the "hosts of heaven," which refers to the multitude of celestial objects and may also refer to the angels where the context of the text makes this evident.

*"Thus the heavens and the earth were completed, **and** all their hosts."*

Genesis 2:1   (NAS)

This should not come as a surprise because it follows that geometric SPACE is the second basic component in the creation of our **TIME, SPACE, AND MATTER,** physical universe.

So, God is not only the author of Time, but also of the concept of spatial geometry.   Interesting - so what then are we being told in this opening verse?   That God created TIME, SPACE, and PLANET EARTH? NO!

God created, along with TIME and SPACE, the *"Erets"* - Hebrew for "MATTER", **not planet Earth** at this point in time. The "erets" of verse 1 (translated as earth) simply refers to the basic components of all matter or what we would call "atoms" today. The awesome truth revealed to us here is in complete correspondence with the current understanding of the makeup of the physical universe. Simply that God is the creator - Ex Nihilo - of the known physical universe and He identifies those fundamental components for us in verse 1 as:

TIME, SPACE, AND MATTER

Therefore, we can now understand that, at this time, the initial phase of creation consisted of only the raw materials. This is confirmed by the statement of VS:2 that follows immediately whereby it is stated:

"*and the erets* (unformed matter) *was without form and void* (uninhabited)."

Hereafter, as the account of creation continues in VS:3, the basic components of time, space, and matter are to be ordered and structured ("framed," Hebrews 11:3) into the planetary bodies and life itself.

Next, I would like you to scan the text of Genesis 1. Did you notice that *every* verse, after the opening revelation of VS:1, starts with the conjunction "And" or "Then"? This is very significant because the Holy Spirit, who inspired the text, has connected all 30 verses together as a unit. The use of "and" or "then" provides an **immediate and sequential connection** (like Super-Glue) to each previous statement or act of God. The insertion of verse numbers was merely added by man for convenience of reference. This pattern should demonstrate to the reader a CONTINUOUS SEQUENCE OF THE CREATIVE ACTS of God, starting with VS:2. A few translations have omitted, in error, the conjunction "And" at the beginning of VS:2. The original Hebrew text requires its presence, however. Observance of this distinct pattern alone should convey to you that the condition described in VS:2 **follows immediately**, simply as a further description of the existing state of the creative act of VS:1. Therefore, there is no reason to place a break in time between VS:1 and VS:2 any more than there would be for any of the remaining creative acts throughout the Chapter, which are equally connected by "And" or "Then." Thus, the magnificent unified statement of VS 1 & 2 opens to us the revelation of creation as follows:

In the beginning (of TIME)
God created (from nothing physical) **Space and Matter and the Matter** (unarranged elements) **was without form and void** (without life, uninhabited) **at this time.**

The Gap Theory - *Is This Just a Side Issue?*

At this point in your reading, you may be wondering if it is really important to understand the opening chapter of the Bible. It certainly is, because much of the truth and relevance of origins is supported by the foundation of Genesis. With this in mind, let's examine the popular teaching that I believe has adversely influenced multitudes of people as well as the Church today.

Many Christians are taught that there exists a huge time gap between the first two verses of Genesis in Chapter 1. This idea has been introduced even though there is no reason to disconnect verse 2 from verse 1 any more than you would have reason to disconnect the remaining 29 verses, which are also sequentially connected by *and* or *then*.

Genesis 1:1 In the beginning God created the heavens and the earth (NAS**)**

Time Gap Incorporating ⬍ all geologic eras

Genesis1:2 And the earth was formless and void and darkness was over the surface of the deep; and the spirit of God was moving over the surface of the waters. (NAS)

The proposed Gap or Ruin-Recreation **theory** has several elements that may be summarized as follows:

1) Verse one reveals a **completed** perfect creation of the universe which includes the Sun, stars and planet Earth at this time.

2) Satan was initially ruler of the Earth, which was populated by a pre-Adamic race of some kind?? Satan eventually rebelled by desiring to become like God.

3) Sin entered the world because of Satan's desire to usurp God's throne. Eventually, after an indefinitely long period of time, this rebellion resulted in God destroying the Earth with a flood. The waters of verse two, it is alleged, are presumed to be left from Lucifer's flood. Furthermore, the fossils that we collect today, like dinosaurs, are the remains and evidence buried in this flood.

4) The remainder of Chapter 1, starting with verse 3, is therefore proposed as a **recreation** of the earth and all current living organisms.

It is interesting to note that the proponents of the Gap theory have almost uniformly appealed to it for the purpose of accommodating the vast periods of time required by the theory of evolution and accepted by many geologists. The proposed billions of years of 'evolutionary' time is conveniently inserted in the "gap" between verses 1 & 2. Justification for this is not based on translation of scripture. It is **reasoned** that such a time break is needed in order to harmonize scripture with current scientific theories. Unfortunately, those who advocate this theory are unaware of the accumulating scientific evidence that precludes evolution and supports the biblical account of a young Earth. Further details of such evidence are provided in later chapters of this book. It is not co-incidence that the raw data of factual scientific evidence corresponds to the Word of God because the handiwork of God, seen in the physical earth, will not contradict the mind of God - revealed by scripture! (Romans 1:20)

The ruin - reconstruction ideas were primarily introduced by the Scottish Theologian, Dr. Thomas Chalmers in the early 1800's. This man was a contemporary of Charles Darwin and Charles Lyell, who came to be known as the founder of uniformitarian (long ages) geology. Think about this: is it possibly just a coincidence that the Gap theory happened to be proposed at the same time that geologists were introducing extreme ages for the earth and evolution was being promoted as worthy of scientific status? Why is it that such a division in the opening verses of the bible was not detected centuries earlier by hundreds of scholars or early church leaders?

Is it really prudent to attempt to harmonize scripture with current unfounded theories of science? We know, from experience, that science and its theories have always been changing but the Word of God never changes!

"Forever, O Lord, your Word is settled in Heaven."

Psalms 119:89

Actually, the preponderance of evidence being accrued by scientists today has now revealed the fatal deficiencies of evolution and concepts of an old earth. In light of the relatively recent discoveries disclosed through the current Information Age, no rational basis exists any longer for such a belief. Therefore, it is **not even necessary or logical** to propose a huge time gap between verses 1 & 2. For example, scientists have been continuously changing their estimates of the age of the Earth. Between 1900 and 1960, the proposed age of the earth increased from 50 million to 4.5 billion years. Importantly, it is not common knowledge that the reported ages of the universe, as well as the solar system, the Sun, the planets and the moon are **all dependant upon** the assumed age of the Earth.[#1] Significantly, most scientists are not aware that these ages have not been verified by any independent measurements and the assigned ages have periodically been changed as the direct result of revising the age of the Earth. The magnitude of the importance of this issue cannot be overstated and is disclosed in the reference cited by Dr. Jonathan Henry; summarized as follows:

> "The long chronologies for the universe and its parts are therefore not independent of the alleged old age of the earth. If the earth is shown to be young, the evidence for an old universe crumbles"

The current published age of the Earth now stands at 4.6 billion years, however, this perceived age is primarily based upon the techniques of radioisotope dating. This method relies upon the highly questionable application of applying theoretical assumptions, which **have been demonstrated** to yield inconsistent erroneous results. The generally accepted dates, however, continue to be published as factual, even though numerous scientific evidences have been established which point to a relatively young Earth.

Please be aware that God is not just standing by as all this deception blankets the Earth. The good news is that the Holy Spirit is raising up many Creation Science oriented ministries. These local and international ministries work closely with thousands of scientists worldwide to delineate the direct correspondence between the current state of scientific discoveries and scripture. The truth is now available to "whosoever" will receive it. Never before has the evidence for creation been so readily available in multi-media seminars, complete with displays, as well as books, and video tapes. Issues concerning origins are really easy to understand and the truth is being widely circulated in these last days in a non-technical format. After all, the truth is usually simple!

Correct understanding is extremely important because the issues of origins reflect upon the integrity and credulity of Genesis - the **foundation** of God's Word. If the foundation is rationalized as flawed and essentially "cracked", how can it be used to support the rest of the Bible?

"If the foundations are destroyed, what can the righteous do?"

Psalms 11: 3 (NAS)

Compromise By Intimidation

Consider this: Those who promote the gap theory are apparently not aware that, if this interpretation of the Bible could make room for a vast expanse of time, then the entire creation account becomes progressively inconsistent with the tenants of evolution. This is because:

BIBLICAL CREATION OR EVEN A RE-CREATION ARE TOTALLY INCOMPATIBLE WITH EVOLUTION

NOTE: the cryptic message of evolution is that there is no Creator!

When a theologian attempts to harmonize evolution with scripture, he is actually contradicting both the fundamentals of evolution theory as well as the clear teaching of scripture itself.

For this reason the evolutionary community **does not accept** the biblical account of creation even through an extended time frame may be "slipped" into the Genesis creation account.

The Tables that follow will make this immediately evident:

CHECK THIS OUT CAREFULLY

THE VERY NATURE OF GOD IS TOTALLY INCOMPATIBLE WITH THE CONCEPTS OF EVOLUTION

SCRIPTURE √	THEORY √
Divine Creation	Life From Non-Life
Purposeful Design	Random Accidental Order
God's Word Creates	Time Creates
Entropy (decreasing order)	Evolution (increasing order)
God's Plans & Purposes	No Purpose
God's Standards	No Standards
Man Degenerating	Man Improving
Death Due To Sin Of Man	Death Prior To Man
A Future Hope	Hopelessness
Eternal God	Eternal Matter

LOGICAL CONCLUSION...

SCRIPTURE & EVOLUTION SIMPLY CANNOT BE MERGED, LINKED, RECONCILED, HARMONIZED OR MIXED IN ANY CONCEIVABLE MANNER

Are Creation And Evolution Really Compatible?

Consider this further…

When a Christian minister attempts to merge evolution or long age concepts into scripture with the intent of preaching the Gospel to those who are inclined to believe evolution, he obviously does not realize that a person knowledgeable in evolution theory will recognize the inconsistencies and reject the attempt.

The following Table clearly delineates these additional major inconsistencies of theistic evolution. (The belief that God used evolution as a mechanism to create life)

ORIGIN OF THE PHYSICAL UNIVERSE	
Biblical Account **CREATION**	**Theistic Evolution** **BIG BANG**
Earth, then light, then stars	Light, then stars, then Earth
All life created fully formed and complex	Life started from non-living chemicals and increased in complexity
Man created in the image of God and given dominion over animals.	Man descended from Apes.
The Earth is about 6000 years old.	The Earth is about 4,500,000,000 years old.
The genealogies from Adam to Joseph are mathematically linked by the patriarch's age at	The Biblical genealogies from Adam to Joseph are assumed to contain time gaps which may

the time of the next descendant's birth. Therefore, no additional time can be inserted in the human lineage as recorded in scripture.	span centuries.
Exactly 77 generations are specified between Jesus and the creation of man by God. (Luke 3)	Man descended from the common ancestor of apes over a period of about 4,000,000 years. If true, this would require about one hundred thousand generations.
Nearly all the fossils within the rock sediments were entombed rapidly during the world wide flood. Death & destruction was introduced into the world by the sin of Adam. Most fossils are about 4500 years old. Gen3:17 & Rom 5:12)	The fossil record recorded in sedimentary rocks has accumulated over a period of about 600,000,000 years.

Therefore:

BELIEF IN THE GOD OF THE BIBLE AS THE CREATOR OF EVOLUTION IS IRRATIONAL

REFERENCES

1. Henry, Jonathan F., 2003, *An Old Age for the Earth Is the Heart of Evolution,* Creation Research Society Quarterly, 40(3): pp.164-172

Two Ways To Know The Truth

ORIGINS- 21st CENTURY UPDATE

Security Zone Briefing

The intent of this chapter is to present a candid update on the major issues concerning the origin of life on planet Earth. The reader will then be able to make a more knowledgeable decision regarding the 'correct choice'.

When humanity entered the 21st century just a few years ago, emotions ran high due to the uncertainties of a rapidly changing world. Y2K was not the only cause for anxiety. Cultural diversification, political correctness, job market changes, workforce reductions, religious tolerance and church vs. state issues rolled in upon us like a fast moving cloud. The Information Age has ushered in technological and cultural changes faster than many souls could adapt to. (except for teenagers-of course!) As a result, the rapid pace has challenged the economic, social and spiritual comfort zones of many individuals. Millions of persons can now communicate world wide in an instant. E-mail and cell phones are commonplace. A person can now search all over the globe for employment, sending a resume and applying for a job on-line without ever licking a stamp. Men are tampering with life itself through cloning and genetic engineering of foods. What law abiding citizen of the United States would ever dream of witnessing the removal of a monument or plaque containing the ten commandments from public school grounds or a classroom? Obviously everything, including our traditional cultural values, is changing about as fast as computer technology. After all, no one can seem to own a current computer or software for too long!

O.K. so where are we in our overall knowledge of life's origins? The answer is predictable. The Information Age, along with rapidly advancing technology, has also provided a quantum leap in the understanding of life's mysteries. But as you read on it will become evident that this leap is not in the direction desired by the evolutionary minded community. Everyday people are struggling however, just trying to keep pace with the flood of new data and technology to remain employable. There seems to be very little time available to ponder the issues pertaining to life's origins. As a result, many people tend to feel comfortable by just staying where they are, otherwise they may have to re-evaluate their evolution oriented thinking. Unfortunately, this briefing may upset the security of their comfort zone. For those who are flexible and open to some serious re-thinking, the information that follows is intended to provide busy people with an update that can bring the reader up to pace with the Information Age.

Where Does Life Begin?

Most people just have not given much thought to this question beyond what they are presented through the popular media of T.V. programming. (NOVA – PBS – National Geographic, etc) Does life really begin auto-matically by forming microscopic organisms from dirt? Some people still think so but the long standing concept of **abiogenesis,** whereby life is proposed to form from non-life, cannot be supported by any of the following:

> ➢ scientifically verified evidence
> ➢ observation
> ➢ repeatable empirical experiments
> ➢ principles of genetics
> ➢ mathematical probability
> ➢ natural chemical reactions

The fact is, evolution oriented scientists have been eagerly anticipating the verification of abiogenesis since the publication of Darwin's *"Origin Of The Species"* for about 150 years. However this hopeful expectation has repeatedly been frustrated. High tech advances in research over the last four decades have generated streams of data that oppose this concept. Life starting without a biologic origin has simply remained the dominant world view solely through faith in the **unfounded** belief in a naturalistic origin of life, devoid of a creator. Press on and decide for yourself what you choose to believe.

Abiogenesis is the proposed launch pad of evolution. Its proponents insist that free-living organisms can rise spontaneously from non-living molecules by the random combinations of natural chemical reactions. This theory proposes to start evolution on its long journey of developing all life forms and is portrayed as the absolute foundation and unifying principle of the life sciences.

Going further, the theory is usually presented as established fact and is elevated to a status equivalent to a universal law such as Newton's three laws of motion, gravitation, or the laws of thermodynamics.

The idea of abiogenesis may have appeared reasonable several centuries ago but now every thinking person should consider this important question.

> Has the weight of accumulated evidence
> to date reinforced this foundation or has it
> weakened the supporting structure whereby
> evolution is launched?

The following information should provide a basis for answering this question and drawing your conclusion.

> ➤ You will not find anywhere in the world of scientific publications a demonstration, plausible scientific evidence, or research to support the hypothetical starting point of the evolutionary chain.
> (life from non-life)

Dr. Jerry Bergman (Ph.D. in human biology; & Ph.D. in Evaluation & Research) emphasizes the current status of abiogenesis:

> "No one has proven that a self-copying molecule can self generate a compound such as DNA. Nor has anyone been able to create one in the laboratory or even on paper." [1]

Each year more professional publications accumulate attesting to this status. Advanced research tools have shown us, to our amazement, that the building blocks of life consist of cells that typically contain thousands of diverse complex parts. Like an unsolved mystery, naturalism can not account for these nano parts for they are not imported ready made, but are always manufactured within a pre-existing cell. A cell then is not merely a static warehouse of specialized parts (proteins). It is more analogous to the sophisticated, high tech, secretive complex of "Area 51". Within the confines of this complex, the organic components are first synthesized and then undergo assembly into subunits that perform a multitude of functions equivalent to small specialized machines. Like Area 51, much of the internal operations are not obvious to 'outsiders' so consider this information your personal classified briefing.

A living cell then is essentially an extremely complex miniaturized factory that organizes all the internal functions into a coordinated goal of sustaining life and reproducing a replica of itself. Man with all his wisdom, efforts and tools of modern technology has not remotely approached producing such a marvelous device.

Because this is now evident, no knowledgeable scientist today would classify even the smallest single cell bacterium as being primitive or simple. Such a former notion is an ancient relic of historical interest only.

> " it would be an illusion to think that what
> we are aware of at present is any more

than a fraction of the full extent of biological design. In practically every field of funda-mental biological research ever-increasing levels of design and complexity are being revealed at an ever–accelerating rate"

Dr. M. Denton [2]

Today you can search the internet using the key words microbiology, genetics, mycoplasma or bacteria. It is astounding how much research is being conducted to determine what constitutes the least complex biological cell. Great efforts are expended to discover the smallest entity which is capable of functioning as an independent life form. Heretofore it was assumed that such a life form would be simple enough to somehow permit an evolutionary leap from non life to life, i.e. spontaneous generation. Neither viruses nor parasites can qualify as living cells because they can not exist independent of preexisting cells or organisms.

Apparently, to date, an organism which appears to satisfy this criterion has been tentatively identified. The smallest organism being scrutinized as a candidate for this purpose is the bacterium Mycoplasma genitalium (M.G.). Many researchers are studying this bacterium because it is inferred from M.G. that a completely independent organism can not be less complex. Technically, M. genitalium is actually a parasite, therefore, it actually represents a very conservative **starting point** for the minimum requirements of life. Frazer and others have documented some of its characteristics.[3]

Length: 20 nanometers = 200 (10⁻⁹) m

DNA consists of 580,000 base pairs

482 genes

Even though it would take 50,000 of these microbes end to end to fit in the space between the two lines below, this miniature factory manufactures approximately 600 internal parts (proteins).

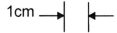

1cm

length of M. genitalium enlarged 50,000 times

Despite its minute size, M.G. still requires 482 genes to replicate. This appears to represent the lower limit of the known genome (gene content) that is capable of independent replication. Below is an image of M. genitalium obtained by an electron microscope from the University of North Carolina. Notice the length of the bar scale is greatly magnified. In reality, this scale is only one millionth of a meter long. (.000001m)

21st Century Update
Electron Microscope Image: Mycoplasma genitalium

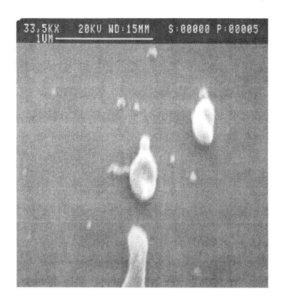

Reprint permission courtesy of:
K. Frantz, A. Albay & K. Bott
University of North Carolina, Chapel Hill

The rapidly advancing field of microbiology has permitted us to peer into the micro cosmic world. From the evidence streaming in we may begin to appreciate that cellular life is a highly organized social event. Inside each cell many subunits are found acting as biologic organs. These nano machines are programmed to operate in accordance with the instructions received from the DNA software. In addition it is obvious that all these parts relate to each other through multitudes of infinitesimal connections. The resulting communication network is analogous to a state of the art web based chat room.

The DNA molecule contains a comprehensive blueprint that facilitates the operations of the cell factory. A few typical operations include food intake, repairs, replication, and defense as well as coordination with many other cells if it is part of a multi-celled organism. Although the marvelous DNA blueprint is complete in itself, it is insufficient to yield life. Have you ever viewed an electrical schematic? Intuitively you know it contains a complete set of instructions, yet a knowledgeable technician is needed to interpret the information and carry out the instructions. Similarly, an operational cell must also incorporate the means for the information content of the DNA to be utilized. This requires that the coded instructions be translated, transmitted and organic assemblies performed by the incredible internal machines which must be **already** in place and integrated into the cell factory.

I imagine that you are now beginning to realize the implication of all this! These complex interdependent operations are way beyond the realm of Darwinian biology.

Now let's consider an analogy to an automobile. The life sustaining processes of a cell factory are parceled out to various 'company' departments. M. genitalium, for example, has 482 departments which we call genes. Each gene is like a file folder on the DNA drive. The gene contains files of unique instructions. These instructions are in the form of units of coded information that are read to assemble a unique string of amino acids (nuts and bolts for our auto). The strands of amino acids are then manipulated by micro packaging units into the form of three dimensional complex protein molecules. This stage is somewhat analogous to manufacturing the sub-assemblies that an automobile requires such as a starter, transmission, battery, generator etc... These components (proteins) comprise the building blocks of life.

Do we have a sports car at this time? Absolutely not! Even if "Mother Nature" could somehow gather all 600 components of the "MG" together, the net result would only amount to a meaningless aggregation of parts stalled on an abandoned assembly line. Now most civilized people can easily differentiate between a collection of auto parts sitting on the shelf and an operational car.

Before we purchase an auto, we demand that it be fully capable of taking us down the road. So also, our current knowledge reveals that the minimum requirement for life **starts** with a fully functional biologic organism. Likewise this demands that life could only start from the end of the assembly line. This is the only place where a completely operational cellular vehicle is ready to operate. But the only assembly line available is within a preexisting cell factory because more sophisticated operations are still needed to complete the assembly of all the parts.

This dilemma has left naturalism without a satisfactory explanation by which all the components of life could be manufactured or assembled.

In light of these newly discovered cellular revelations there has arisen an urgency among evolutionary scholars to propose additional mechanisms in order to maintain credulity. Any such process, of course, must be limited to the natural laws of chemistry. Although natural selection has traditionally been accepted as the primary cause of evolutionary change, nearly all naturalists today will now acknowledge that N.S. is not capable of such a task. The process of natural selection is evident to all but its effects are limited to favoring a trait or organism that exhibits a distinct survival benefit. Thus, N.S. is precluded from promoting the chain of events leading from non-living chemicals to a functional cell because intermediate non-functional stages have no survival benefit.

Dr. Ariel Roth is the author of over 140 articles on origins. He explains why Darwin's mechanism of natural selection cannot result in a functional cell.

> "The problem is that the very system of natural selection which Darwin proposed will tend to eliminate the interdependent parts of complex systems as these systems develop. The parts do not function until all the interdependent parts are present and the system works and provides some survival value to the organism. These non-functioning parts will tend to be **eliminated** by a natural selection process that should give preference to organisms that are not encumbered with extra useless parts."[#4] (emphasis by author)

The need to find another natural process or an additional mechanism eventually lead to the hypothesis that mutations acting in concert with N.S. were sufficient to yield a continuous expanding chain of organic complexity. The addition of this factor is known as **neo-Darwinism**.

As research again progressed, however, the initial confidence in the efficacy of mutations has not turned out to be a breakthrough for evolutionary theory. The science of genetics has established that a mutation will nearly always have the negative effect of producing a DNA copy error or omission of a gene during reproduction. At best a mutated gene may remain dormant for several generations when reproduction utilizes two sets of genes, one set from each parent. This condition cannot result in the advancement of living forms because mutations are actually responsible for **degeneration** of function. It is overwhelmingly clear that degeneration then moves life in the opposite direction from proposed evolutionary advancement. Neither can natural selection help. As pointed out by Dr. Roth and numerous other geneticists and biochemists, a mutant trait or organism will eventually be eliminated by the process of N.S. because this novelty will not have any long term survival benefit.

In response to all these self-evident genetic facts, evolutionists have further proposed that among mutations, a rare mutation may theoretically exist that could possess a potential survival benefit. Fortunately any mutation is a rare genetic event. Therefore the theoretical occurrence of a 'beneficial mutation' is further reduced to an extremely remote possibility.

UPDATE: Numerous independent experiments with
            organisms over literally thousands of gen-
            erations have failed to scientifically demon-
            strate any mutation which could be con-
            sidered as beneficial.

Even so, genetic analyses have been published concerning the probability of the effectiveness of a **hypothetical beneficial mutation**. The question is posed: Could an extremely rare beneficial mutation yield a long-term positive evolutionary effect if it is included during re-production among the much greater normal (negative) mutation rate?

One such study focused this question on the development of the human race from its assumed chimpanzee ancestors. The results indicate that the population of the pre-human forerunners that would have been required to generate the beneficial evolutionary genetic changes through successive reproduction and deaths is totally inconceivable. So again and again, the world of facts procured from modern genetic research virtually eliminates the neo-Darwinian theory based upon natural selection and mutations as a workable hypothesis.

What Do You Know About Dune Buggies?

I imagine that the information update already presented in this section would be reasonably sufficient to make a decision concerning the "two choices". But keep going and the obvious should become overwhelming.

The concept of **irreducible complexity** is logically intuitive but if you are like myself, I never gave it much conscious thought until it was presented in 'black & white'.

Think about any functioning system or machine. Regardless whether the machine is electrical, chemical, mechanical or biochemical – no matter how large or how small- the given machine can only perform its intended function by the coordination of all its necessary inter-related components.

Can this system work with some broken or missing parts?

Now you may be thinking - but I know some machines which can operate even with some broken or missing parts. This may be true until you arrive at the minimum number of

necessary parts. Lets consider again an automobile, for example. Sure you don't need many unnecessary items on your Bentley for it to get you down the road. You can pull out the Bose stereo and take off the doors. How about throwing away the electric window motors but eventually you will arrive at a point whereby every remaining part is absolutely essential to accomplish its primary purpose of basic transportation. It goes without saying that the vehicle requires a starter, wheels, hundreds of engine components, transmission and a multitude of associated interconnections.

What do you suppose would be the minimal configuration of your Bentley in order for it to still perform as an automobile?

Well you could completely strip it down to a dune buggy status but remember even a dune buggy requires hundreds of compatible parts – doesn't it?

So you can also apply the same rationale to living organisms. Anything considered to be 'living' must consist of at least one cell. Even the smallest biological living cell consists of millions of atoms and numerous operational sub-mechanisms.

So again we come to the conclusion that the vehicle of life is essentially a molecular nano machine which, like the dune buggy, must have all its necessary **parts and connections** pre-assembled and in working order right from the beginning. Anything less than the minimum required parts results in an inoperative mechanism.

Now also consider this carefully. Anything less than all the required connections will also result in a useless mechanism. Nearly everyone has experienced the frustration of a totally unusable automobile when it will not drive out of the garage. Why is this? Just because one vital connection among thousands is inoperative. I.E.- one rusty battery connector. Perhaps the engine starts but it cannot transfer power to the transmission due to one broken gear shift link. In any case, the net result is the auto is no better than 3000 pounds of metal, glass and plastic sitting in your garage – basically useless! It cannot perform the intended function.

Well, does this represent an application to real life? Suppose you disassembled an ant or a fish. In either case you sterilized the remains and placed them in a container. All the parts are certainly available but life has ceased without the proper connections. Nor are these parts expected to reorganize themselves. Yes - but given enough time wouldn't all these parts begin to arrange themselves into at least a dune buggy…I mean a simple bacterium like M. genitalium? After all, all the parts are available. Think about this. If that were possible, I imagine you would never open a can of organic ingredients like Campbell's soup and eat it. Why have people been canning and preserving foods for centuries? Because they trust that once all remaining life forms (bacteria) are destroyed, the ingredients are not expected to produce any new forms by spontaneous generation. This leads us back to the opening statement of this chapter: namely, science is based on observation or repeatable experiments and no one has ever observed or documented spontaneous generation. By the way, I like Campbell's soups!

Concerning the parts, published research has repeatedly shown that even the individual parts of life (proteins) have no probability of forming by random chemical combinations regardless of how much time is available. Furthermore, even if it could be imagined that several proteins could form by chance, they could not exist long enough outside the preserving environment of a living cell while they 'wait' for the numerous required other proteins to form and subsequently be assembled. Chemists now assure us that organic parts such as amino acids and proteins will quickly break down (2nd Law again) from the effects of oxidation and dehydration without being quickly incorporated into an operational pre-existing living cell that provides the sustaining protection and maintenance.

The biochemist Michael Behe (1966)[5] has conducted extensive research and has published his findings. He concludes that irreducible complexity must be a result of design in the living world. Applying this principle to living systems, a cell in the initial incomplete stage of evolution would not be a candidate for natural selection because it is not yet able to function and therefore has no survival advantage.

The bottom line of all this is the realization that the streaming inflow of 21st century data is clearly pointing toward intelligent design. Hopefully, those who have been saturated with prior naturalistic inclinations will be able to handle the shock.

It is no longer a mystery why...

THE FIRST LINK IN THE EVOLUTIONARY CHAIN REMAINS MISSING

MODERN SCIENCE REVEALS THAT LIFE FROM NON-LIFE IS BEYOND THE REALM OF POSSIBILITY

Is There A Society Of Planet Earth?

Not long ago biologists, zoologists, and naturalists would think of communities as a collection of individual plants and animals. This restricted thinking has never been advantageous for administrating environmental and conservation programs. Successful fieldwork now concentrates on protecting the mutually shared interrelationships of plants and animals as well as their physical environment.

Because of this elevated awareness, scientists who are involved in conservation programs endeavor to utilize the concept of biodiversity. Many endangered species have been rescued due to the application of knowledge gained through biodiversity research.

Basically, we have learned that both plants and animals of this planet require a collective sharing of services among the variety of populations and species of a thriving ecosystem. All organisms literally serve each other. They depend upon relationships which are **absolutely essential** for the survival of each other within the ecosystem.

The requirement of biodiversity runs deeper than traditional logic had imagined. Even the fundamental necessity for food is now understood to link all life into a direct mutual dependency. Food in all its varieties is universally acknowledged as a requirement for sustaining life but ultimately all ecosystems rely on the **preexisting** photosynthesis of plants to convert the Sun's energy into carbohydrates. These molecules provide the energy source of **all** organisms that are not capable of this amazing conversion. But then again, even the plants are completely dependent upon the nitrogen fixing micro organisms within the soil. In addition, plants need animals for pollination and distribution of seed. This emerging body of facts also defies any satisfactory evolutionary explanation because biodiversity negates the possibility of organisms evolving independently of one another – one at a time.

The Information Age has permitted the exchange of volumes of related research among diverse scientific disciplines. The only logical conclusion is that the cycle of biodiversity is in reality a **closed circle** of life that must have been fully functional at the very start. The direct implication is clear – planet Earth was designed to sustain life.

On an even larger scale, the marvelous global ecosystem of planet Earth is totally dependent upon its unique location within the solar system. Studies have been published which indicate the amazing number of physical and chemical properties of our planet that appear to be absolutely critical for the existence of life. Can all of this be accounted for by chance and fortuitous circumstances? The expectations of an evolutionary process would predict a haphazard development of **independent** life forms. Thus the evolutionary model of origins stands in direct contrast to the expanding body of evidence that apparently points more persuasively toward a unified designed creation.

Analogous to a living cell in which all he parts must be up and running simultaneously, so also on the grand ecological scale, our growing understanding draws many scientists to conclude that the complexity of biodiversity appears certain to be a requirement for sustaining all life. The symbiotic interactions that permit a community of plants and animals to survive must, of necessity, come into existence simultaneously...perhaps in the beginning?

Dr. Henry Zuill, professor of biology at Union College, Lincoln, Nebraska sums up the origin of biodiversity as follows [6]:

> "HOW COULD MULTIPLE ORGANISMS
> HAVE ONCE LIVED INDEPENDENTLY
> OF SERVICES THEY NOW REQUIRE?"

Probably Yes or Probably No?

Even mathematics can help us understand about origins. But perhaps you are thinking – I'm not a mathematician. No problem! Not everyone is mathematically inclined but this briefing will provide the curious person with an overview of just what the numbers tell us. And for those who use math as a language, the references cited will lead the analytical types to a place of checking out more comprehensive analyses.

Some scientists who are actively involved in origins research have published calculations that assess the probability of a biological mechanism or even just a single protein forming by the naturalistic process of random chance. Evolution is believed to utilize the existing chemical properties of matter by which atoms combine to form larger organic molecules. In addition, it is conceived that the products of chemical reactions could also be sorted by natural selection and the more sophisticated biological entities, possessing the ability to replicate, could be further influenced by mutations.

Based on these assumptions, a researcher will invariably use an extremely conservative set of mathematical conditions in order to favor the odds of a successful chemical evolution scenario. Several of these interesting inquiries are summarized below in an abbreviated format. Keep in mind that most of these examples only consider individual parts of a cell. Of course, as you have learned, even a box of proper parts does not come close to yielding self-sustaining life.

The numbers expressed in this section are usually expressed as ten raised to an exponent. It is easier to write very large numbers this way. The number (10^{10}), for example, simply represents one followed by ten zeros, i.e. 10,000,000,000.

A few numbers are given for reference:

Diameter of the visible universe = 10^{28} inches

Estimated number of electrons in the universe = 10^{80}

Probability Analysis #1

Enzymes are proteins which perform specific functions within a living cell. Each specific protein is constructed from a unique combination of amino acids which are initially chemically attached end to end in the form of a string. The unique sequence and geometric shape of these building blocks determines the function of the protein. Apparently there are no chemical laws or natural physical properties that can induce any particular sequence. Therefore, the use of mathematics provides a valid application for determining the probability of a particular sequence occurring by chance. The probability of sequencing a typical set of enzymes that are found in a living cell has been examined by Fred Hoyle, a renowned evolution oriented astronomer. He has published his findings in the reference cited. One of his particular conclusions follows – "the chance of random shuffling of amino acids producing a workable set of enzymes"

less than one chance in $10^{40,000}$ [#7]

The significance of this infinitesimally small probability translates to mean in reality -"not possible".

In a prior publication, Hoyle and Wickramasinghe had this to say: "Any theory with a probability of being correct that is larger than one part in $10^{40,000}$ must be judged superior to random shuffling (of evolution). The theory that life was assembled by intelligence has, we believe, a probability vastly higher than one part in $10^{40,000}$ of being the correct explanation of the many curious facts discussed in preceding chapters. Indeed, such a theory is so obvious that one wonders why it is not widely accepted as being self-evident. The reasons are psychological rather than scientific". [#8]

Probability Analysis #2

Has the Earth been hanging around long enough for the minimum number of proteins required by the smallest organism to be formed by chance? James Coppedge has helped to answer this question by estimating the number of years required to obtain the 239 proteins. His results indicate that this event may happen once in $10^{119,879}$ years. Now the proposed age of the Earth based on the evolutionary time scale is 4.6 (10^9) years. So, of course, the time required is way beyond the assumed age of the Earth. This places this singular occurrence totally beyond practicality.[#9]

Probability Analysis #3

Dwain Ford, Ph.D. Chemistry, has examined the probability of forming one coded protein in the simplest known organism of Mycoplasma genitalium. This protein consists of 347 amino acids. The trick for Mother Nature is to assemble the amino acids by random shuffling in the required sequence. He concludes mathematically that it could be done after 10^{451} random attempts. Even though this represents a highly improbable event, it turns out that there does not exist a sufficient quantity of carbon on planet Earth to randomly combine and supply this improbable number of amino acid chains.[#10] Furthermore, the real world presents an infinitely greater evolutionary barrier because M.G. contains 470 genes each requiring a specific protein.

Probability Analysis #4

The estimated probability of a single basic protein molecule assembling by chance [11]:   1 chance in 10^{43}

Concerning "Probably Yes or Probably No", the informative comments of two other scientists are provided below.

"We now have seen that self-replicating systems capable of Darwinian evolution appear too complex to have arisen suddenly from a prebiotic soup. This conclusion applies to both nucleic acid systems and hypothetical protein-based genetic systems" [12]

"However, the macromolecule-to-cell transition is a jump of fantastic dimensions, which lies beyond the range of testable hypothesis. In this area all is conjecture. The available facts do not provide a basis for postulating that cells arose on this planet" [13]

Pilate, the Roman procurator of Judea, asked this question long ago: "What is truth?" Jesus actually did not reply to this foundational question. Why? Because God has made manifest provision for every person on Earth to determine TRUTH for themselves. True science concerns that which anyone can observe and repeat and thus conclude what is truth for themselves. What therefore has all mankind actually observed and repeated countless times since 'Day 1'?

LIFE ONLY ORIGINATES FROM LIFE

It follows that this fact can be relied upon as representing truth and is called the LAW OF BIOGENESIS.

No one has ever documented any other origin for life other than **observing** each living offspring or plant coming forth from the parents or seeds of the same kind. It is this reality that is proclaimed in the Bible's account of creation.

> Then God said, "Let the earth bring forth living creatures after their kind"… **Genesis 1:24** (NAS)

Does Intelligence + Information $\rightleftarrows$ Life

As you read this text right now, you may not realize it but you are actively involved with the basic entity of all life which is: INFORMATION.

Our physical world incorporates countless chemical combinations and disintegrations each day. Remember how reversible chemical reactions are indicated?

$$A + B \rightleftarrows AB$$

These symbols indicate that the chemical reaction can proceed in either direction.

For sure, living organisms include many such physical reactions but life also requires an analogous Non-Material counterpart that is every bit as real. This 'virtual' reaction may be symbolized as follows:

$$\text{Intelligence + Information} \rightleftarrows \text{Life}$$

Even scientists have difficulty defining life in physical terms because it is characterized by the intangible entities of intelligence and information.

> The existence of information in our universe stands as an absolute barrier to evolution and all its tenants.

This bold statement may initially seem to be exaggerated or even extreme but a few moments of information transfer will allow you to appreciate why information is really impossible to account for, from any naturalistic standpoint of origins. The revelation that life is a by-product of information may seem foreign to many who have not considered this concept. Be assured, it is relatively easy to understand.

Up to this point many people comprehend that the physical universe consists of the fundamental entities of:

TIME - SPACE - MATTER

We also know that matter & energy are closely related and interchangeable:

$$MATTER \rightleftarrows ENERGY$$

So also, it appears that time & space are related in such a way that time itself may slow down or speed up relative to the reference frame of measurement.

$$TIME \rightleftarrows SPACE$$

Traditional Concept of The Universe

TIME - SPACE - MATTER

Scientists are now coming to grips with the fact that one additional entity is also an integral component of the universe: namely INFORMATION.

Heretofore, this component has been basically overlooked perhaps because it has just been taken for granted and the obvious has not been given much consideration. However, traditional thinking changed dramatically when the relevance of information to life's processes was amplified through a significant discovery in 1953.

First of all, it is necessary to understand that information is completely distinct from and independent of time - space and matter. Yet its existence is so uniquely profound that it is now recognized as a fundamental component of our universe. Without information, planet Earth would be devoid of life. Perhaps this is why the analyses of over a ton of samples from the Moon and two unmanned probes on Mars could not detect any life!

During the late 1960's NASA launched the first manned missions to the moon using the Apollo space vehicles. Scientists were initially very concerned about the potential contamination of the returning astronauts with pathogenic microbes. It was envisioned that organisms may have evolved which are foreign to any on Earth. As a result of this reasoning, the first several missions required the astronaut crew to be quarantined in an isolated chamber immediately upon arrival from the Moon. This practice was eventually abandoned as unnecessary after it was determined that neither the space crew nor the samples contained any micro-organisms. It became obvious that no life forms had evolved throughout the proposed four billion years of the moon's existence. Apparently, for some unexplained reason the neo-Darwinian process of natural selection and mutations, given eons of time, has failed to yield any life forms on our planetary neighbors of the Moon and Mars even though it is recognized that Mars had an abundance of water on the planet in the past.

The expanding body of knowledge provided by the Information Age has revealed that the recipe for yielding "life" itself must contain more than the physical parameters of time, space and matter. The quality of existence that we characterize as life requires that the three physical components must be organized by the absolutely essential fourth ingredient of information. Thus **the contemporary concept of the universe has now been updated** to reflect our current understanding as illustrated below. And so you may be asking yourself, 'what was the ultimate source of this fourth ingredient'? Does not the very first opening verse of the Bible reveal that in the very beginning of creation **God** was already there?

In the beginning God created the heavens and the earth

Genesis 1:1 (NAS)

Now consider this amazing scripture:

In the beginning was the Word, and the Word was with God, and the word was God   **Jn 1:1** (NAS)

Think about it - are not words the smallest packets of intelligent information?

Contemporary Understanding of the Universe

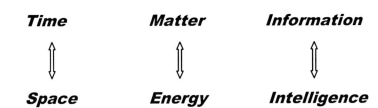

TIME - SPACE - MATTER - INFORMATION

Time	*Matter*	*Information*
⇕	⇕	⇕
Space	*Energy*	*Intelligence*

Information is a common part of our everyday lives. It is encountered everywhere we turn but you can't touch it or feel it. Yet we know it is a vital reality. Well what is information anyway? Obviously information does not consist of chemicals (matter) nor can it be measured like a physical property, or as energy. **Information consists of ideas, instructions, or thoughts.** Therefore, just like Time & Space:

Information is non-material

The CD has no additional weight after 700 Mega-bytes of coded information is stored on the disk.

As such, information can be transferred or copied from one medium to another but the medium upon which the information is stored is completely independent from the information content itself.

As an example, many different mediums can be used to convey the same information. Right now you are receiving information that is stored on paper and ink. This information could just have well been e-mailed to you or have been conveyed through the various mediums of light signals, wireless cell phone (radio waves), hardwire (Morse code/telephone), floppy disk, CD-ROM, or sound waves.

131

Note carefully that the paper and ink of this book is not information any more than smudged ink or random letters on a page would be. The pages of this book are only a storage bank for the information that you are receiving right at this moment. Did you just catch the clue to understanding the relevance of all this? The key is **you** but it also includes **me**, the author.

By now I imagine you are catching on to the concept. Information is not the tangible book you are holding but rather the thoughts and ideas which are being

$$\text{Relayed} \quad \rightleftarrows \quad \text{Received}$$

through **origination and translation** of this printed code which we call language. Thus, the absolute prerequisite for information is the existence of an intelligence having the ability to originate or translate and interpret. Think about it – doesn't information always involve a sender and a receiver?

Dr. Werner Gitt currently directs the German Federal Institute of Physics and Technology. He has published extensively in the emergent field of Information Science. The following statement would normally be considered self-evident but it is also supported by years of his detailed research.

> ## NO INFORMATION EXISTS
> ## WITHOUT AN INITIAL MENTAL SOURCE [#14]

Although information is intangible, nevertheless its reality represents a distinct entity of the known universe. Its existence is obvious but it is neither time, space or matter, yet it is directly related in that realm of the universe involved in life. O.K. but how does this relate to understanding origins?

The obvious direct connection between information and all life was somewhat veiled until a remarkable discovery was made in the field of microbiology. In 1953 the structure and chemical properties of the fascinating molecule known as DNA was revealed by Francis H. Crick and James D. Watson. The DNA molecule is usually located in the nucleus of a cell. From just the standpoint of organic chemistry, this research was a breakthrough in advancing the knowledge of cell chemistry. Within a short time the implications of this discovery rocketed way beyond the initial revelation. It soon became obvious that the function of the DNA molecule is to store vast amounts of coded genetic information in the form of a chemical alphabet. The physical structure and chemical properties of DNA simply comprise a medium upon which the code of life was stored. Furthermore, the twisted double stranded molecule is now recognized as the most efficient and compact data-information storage medium known. Various conservative estimates for the storage capacity of a typical DNA molecule range from 500 to 3000 encyclopedic books. This may be analogous to about 30 complete sets of the encyclopedia Britannica.

Like the hundreds of languages which are in use today to convey information, each can be considered to be a sophisticated code. A code must be constructed from a set of rules that are required to translate and convert information from one useful form to another. A program utilizes coded information to direct a planned sequence of tasks to accomplish a goal. A program, in turn, represents a **premeditated** designed application of information for a specific purpose. Perhaps that's why you and I must pay Microsoft big $$ for their software programs. Someone put a lot of thought and effort into creating a specific computer program. Now if you were caught with a pirated program it would be a 'hard sell' to convince someone that it arrived on your computer as a product of accident!

Likewise, the genetic instructions encoded on a DNA molecule function as a complex programmed language. The purpose is similar to software that directs the operations of a computer. Any language, however, can only issue meaningful information if it can be translated by someone or by a device that has the capacity to utilize coded instructions. This is accomplished in a biological cell by an elaborate series of incredible decoding operations, usually performed by specialized proteins called enzymes. In addition to these amazing operations, each cell comes complete with its own 'spell check' editor. As the genetic code of life is replicated and passed on from cell to cell, researchers have uncovered several editing mechanisms that protect against transmission errors. Fortunately, this is one reason why mutations (copy errors) are rare. New details, unveiling the mysteries of cellular operations directed by information transfer and translation, are continuously being discovered to the utter amazement of the scientific and medical community.

There appears to be no end to the intricate multiple functions required to sustain life.

The logical bottom line is this:

It is now quite evident that the basic physical chemical component of life (Dirt) requires a multitude of coordinated cell activities which in turn must be directed by a vast information content.

So where does this leave evolutionary theory? If ever it was reasonable to think that neo-Darwinian mechanisms could substantially account for the existence of life, there is no longer adequate justification. When compared to the 'gigabytes' of programmed instructions found within DNA, the proposed evolutionary mechanisms are extremely crude. The mechanisms of natural selection and mutations fall far short of being able to draft the schematic of life, much less directing construction or sustaining the manufacturing plant now recognized as the micro universe of the living cell. The totality of contemporary research does not leave any room for a fortuitous naturalistic explanation for the origin of life-producing information nor life-sustaining information which is an absolute necessity for the perpetuation of life's functions.

Any satisfactory naturalistic model must not only demonstrate a mechanism which can build its own hardware but also originate, download and install a complete operating system (far superior to Windows XP) which can manage a complex cell factory. All such attempts so far have been merely unsupported fantasy and stretch one's credulity in light of the current state of technological comprehension of life's processes.

> A 'Cell Factory' can not open for business
> unless all the organic hardware and internal operations
> can be successfully 'booted up' together
> by the information software

The expanding field of Information Science has shown that a random series of symbols cannot ever develop into a set of programmed instructions (i.e.- language) through proposed neo-Darwinian mechanism of natural selection and mutations. Even a pre-existing language that is copied with continuous grammatical (genetic) mistakes will become unintelligible in short order. Therefore, there is no logical or scientific reason to expect a language or program to arise from the proposed initial chaos of a universe starting from a state of unordered, unformed molecules. Again, such an initial state as well as the subsequent 'self-organization' is also precluded by the 1st and 2ⁿᵈ Laws of thermodynamics. The utter inability of proposed evolutionary mechanisms to account for organically transmitted streams of information is a fatal deficiency of this materialistic scheme.

It is currently recognized that even the smallest units of living things such as M. genitalium possesses enormous quantities of instructions. Because no amount of Darwinian philosophy can account for the chance occurrence of genetic codes, **informed** scientists are left with basically two choices.

1. Those who agree that information could not be formed by any proposed evolutionary process and thus acknowledge that life is a product of creation.

2. Those who have no explanation for the origin of cellular information content but continue to believe in the evolution of life scenario based solely on philosophical faith.

But what about the non-scientist like me? Yes there are still many persons, some even scientists, who have not kept up with the Information Age or who are just misinformed. Unfortunately, they continue to believe that neo-Darwinian factors can explain the origin and development of life.

Each person is at liberty, of course, to place his faith in a belief that chemicals can combine to reproduce themselves and somehow originate and increase their own information content. This would actually require that 'Dirt' could think for itself and direct its own evolutionary development in order to obtain a foreseeable survival benefit. Apparently such a willful faith can only be exercised by overriding objective science. But now you have been briefed and it is hoped that your security zone has been reinforced rather than disturbed.

So, this brings us to the stalled evolutionary launch pad again. The marvelous discoveries of programmed organic nano machines reaffirm the necessity of a launch vehicle which non-living dirt must possess in order to lift off into the journey of life.

The Future of Mankind

This concluding briefing is herein labeled as "sensitive". The update which follows may not be considered politically correct, popular, or newsworthy and could be somewhat disconcerting or possibly even disturbing. Be assured you will not find this information in popular magazines, on the 6 o'clock news, in newspapers or even in school text books. It simply doesn't make for "good press" Why? Because it represents the antithesis of evolutionary dogma which is being promoted everyday everywhere as factual.

The support for this information is available, however, in the Bible as well as professional scientific publications, including creation science journals and related books.

First, a brief summary is presented of the anticipated near future from the viewpoint of the collective thinking of the evolutionary establishment. The mindset of many in this world is based on the ideology of an undirected, purposeless, imaginary mechanism of Darwinian evolution. This paradigm is so entrenched in the world of academia that any student, teacher, scholar, or university professor is severely intimidated from even questioning this granite monolith of modern science.

This entire monument rests on a single faith based foundation:

> The human race is simply an end
> product of accumulated mutations
> which have been sorted out over
> time by the omnipotent power of
> natural selection.

MUTATIONS + NATURAL SELECTION ⇌ LIFE

If you were to take the time to Google "the future of Homo sapiens", your computer would be flooded with numerous speculative pseudo-scientific articles. These dissertations attempt to prophesy the direction and future condition of the human race.

Now because nearly all the prognosticators are evolutionary oriented, their forecast will invariably project the potential beneficial advancement of mankind. Most articles will try to convince you that we are destined to evolve into super-humans through the agency of natural selection, beneficial mutations and the intervention of technological man.

The possibilities which are envisioned are numerous, with the hope that through natural selection an inexorable trend toward a more complex and greater quality of life is assured. Many of these articles confidently assert that our divergence from apes occurred about five million years ago and we are engaged in an ongoing upward progression. It is further believed that the human species (Homo sapiens) emerged in our present form about 100,000 years ago. We are assured that such a long data base of evolutionary progression surely justifies extrapolation into the future. Some writers believe that modern human evolution is accelerating and may be expected to produce major changes in the near future.

A number of visionaries have also considered the possibility of human extinction. After all, the fossil record records that numerous species have ceased to exist after a

140

few million years of survival of the fittest! In lieu of complete extinction, it is speculated that a new species of "Survivalistians" may spring forth as a remnant of post apocalypse survivors. This unthinkable scenario begs the question; are humans so unique from other species that they may be able to avoid any number of extinction potentials by manipulating their destiny? Repeatedly it is reasoned that extinction due to some catastrophe or major environmental change is unlikely because man is now a master of mechanical and medical technology.

Basically it appears that the consensus of speculation on the future of the human race is optimistic. However this utopia may require the enactment of a globally sanctioned policy which mandates the **control** of human evolution by manipulating the fitness of the population. Regulation is a common theme among these evolutionary scenarios as it is believed that it may be necessary to advance a globally blended genome of "Unihumans". Uhm... sound familiar?

Other writers envision the following in the near future:
"Numans"   (genetically enhanced people)
"Cyborgs" (human /machine hybrids)
"Astrans" (genetically optimized for interstellar colonization)

Lets pause and think for a moment... Nearly everyone ponders about where this evolutionary process may take us in the near future. Can we really place our hope for human development in such a process? But at this point you have been exposed to a preponderance of information which clearly compels a re-thinking of the whole evolutionary myth. Do you still believe that the basic evolutionary premise of onward and upward – ever improving naturalistic advancement is really valid?

What follows in this briefing is a revelation of the inevitable future trend of the human race if left solely to its own natural consequences. This is not based on a prognostication of future visions. This insight is based upon the reality of accurate historical data and current state of the art empirical genetic research.

Having established for all practical purposes the impossibility of the spontaneous generation of life from 'dirt', it is now only reasonable to conclude that life is solely the product of creation by God. In fact, God has graciously left us a historical record of life right from the beginning.

Numerous scholars have consistently demonstrated the meticulous transmission and accuracy of the biblical record. These scholars are not limited to theologians but include many acknowledged scientists. For example, most people are not aware that Sir Isaac Newton, considered one of the most remarkable scientist who ever lived, endorsed the accuracy of the Old Testament chronology.[#15] This amazing account of human activity is a gift to us for the enlightenment of mankind and provides us with the fundamental answers which everyone desires to know.

How and when did all this start?

What is my purpose for existence?

*Why is much of humanity subject
to the deplorable conditions of famine,
poverty, disease and senseless
atrocities of war?*

Where are we going?

Science, of course, CAN NOT provide a satisfactory answer to these issues which are essential for every individual to settle in their souls. Scientific investigations are confined to that which can be physically measured. Historical events and issues of life's origins can not be analyzed by microscopes and Bunsen burners.

But mankind has been privileged to have in their position a unique document which reveals the "What", "Why" and "Whens". This unique document is called "The Word of God" and is commonly known as the Bible. It is a gift to mankind by Jesus Christ Himself Who is the Living Word and the first cause of all information.

> *In the beginning was the Word, and the Word was
> with God, and the Word was God.* **John 1:1** (NAS)

Without this document to tell us just what happened in the beginning and to preserve an accurate outline of ancient history, all of us would remain in darkness without reasonable answers to satisfy our God given intellect.

Respectfully be advised that there is no such thing as pre-historic events. The true record starts right at "The Beginning" with the marvelous revelation of creation.

Therefore it provides the basis for logical answers to life's mysteries which would otherwise be unattainable. Keep reading and you will shortly discover the eternal ultimate destiny of mankind. There is no requirement for guesswork here. Humanity has complete access to its ultimate future. Just study the "Word" carefully – its all revealed from the Beginning to the End.

It is not the intent of this security briefing to answer all these questions. You have the actual written account available to you along with an uncountable number of supplementary materials to help explain what God has revealed. But what follows is a condensed explanation of where man has come from and where humanity is headed until Jesus Christ, Himself, the creator and sustainer of this physical universe **intervenes**. For sure, it is the intention of this briefing to assure you that:

> A SUPERNATURAL INTERVENTION INTO THE AFFAIRS OF MANKIND WILL BE ABSOLUTELY NECESSARY IN THE NEAR FUTURE.

Because this is a concise briefing on a subject of enormous magnitude, it is presented as an outline combined with brief descriptions.

A. The beginning of time as we know it started about 6000 yrs ago with the creation of the physical universe and all life forms by the omnipotence and omniscience of God [16].

B. Man was God's supreme creation. He held nothing back because Love is completely generous. Therefore man was created no less than in the image of God Himself and was given **dominion** over the entire creation. His surrounding environment was perfect and included not only complete provision for abundant life but also Eternal Life. Eternal life is a default of being intimately united with God through His Eternal Spirit.

Man was in harmony with all life forms, even the dinosaurs, as there was no need for animals to be carnivores.

> *And God saw all that He had made, and behold,*
> *It was very good.*   **Genesis 1:31** (NAS)

Furthermore:

C. Adam and Eve were given the privilege of close fellowship with God as well as the God like ability to reproduce life in their image also as their offspring. Isn't it obvious that every child born to this day is essentially a creative miracle? Who can rationally deny this?

D. Because Adam and Eve are the progenitors of all

humanity, having been created in the image of God, they must of necessity possess a completely free will. This god-like characteristic remains the awesome gift of every person to this day.

E. Apparently, before Adam and Eve could acquire an intimate knowledge of their loving creator, they freely relinquished their union with God. This decision was a direct result of choosing to doubt God's word to be their total provision. The decision to walk by their own reasoning based solely on their physical senses resulted in grave con-sequences for all humanity. By relinquishing their union with God their connection to Eternal Life was severed, having defaulted their dominion to Satan– a rebellious angelic creation

F. This catastrophe can not be underestimated as it brought havoc to the physical Earth. Thus there ensued the reduction in life span through death. No man was ever born to live apart from God's loving character and marvelous provision. All that has been ever required for any man is simply to **trust** his creator.

With this framework in place –
Now lets look at what the Bible record assures us concerning the actual history of man from the very

beginning. The historicity of the Bible is directly related to the issues of creation, science, age of the Earth and the future of mankind.

> *In the day that you no longer trust Me and decide*
> *to act solely according to your own reasoning –*
> *you will surely die.* **Genesis 2:17**

The translation of the original text literally states "in dying you shall die" Man was never created to live apart from God. The image of God which man possessed at creation was His Eternal Spirit.

> *"the breath of God".* **Genesis 2: 7**

The loss of the image of God is revealed as the first death (spiritual death). Consequently, the second death through physical deterioration was inevitable.

In light of this revelation, it is not difficult to comprehend that prior to the worldwide Noahic flood mankind enjoyed a long life span in excess of 900 years. From our 21st century viewpoint this may seem to be fantasy but even 900 years is in reality a traumatic abbreviated lifespan when compared to God's intent of Eternal Life.

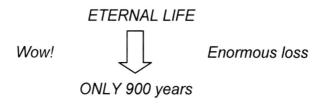

ETERNAL LIFE

Wow!   *Enormous loss*

ONLY 900 years

Aging of the human body has always been a subject of intense scientific interest and scrutiny. Nevertheless, the phenomenon has remained a puzzle. Why should we age in light of the evidence that all of the 100 trillion cells in the human body appear to be replaced at intervals of no longer than 10 years. For example, our epidermal skin cells are replaced every month. Some internal cells replace themselves every four days; red blood cells about every 90 days. In fact, in some sense it could be stated that a senior citizen is really no older than the body of a child.

Indeed we may wonder why pre-flood man only lived about 1000 years. It would appear that normal living cells should divide and grow indefinitely with proper nourishment. However it is obvious that cell division results in a reduction of quality and therefore is limited to a finite number of generations.

Chemists, micro biologists and geneticists are just beginning to unravel some clues which may contribute to understanding the mechanisms of aging. Growing old is a complex process based on many factors. Even so premature aging has been impeded somewhat through advances in modern medicine – at least for those to whom it is available. Conventional medical treatment is limited in scope however. Its greatest potential to prolong life is in the arena of combating malnutrition, diseases and some negative physiological effects related to the environment. But science has not made much progress in solving the fundamental cause of aging which is now understood to be directly related to genetic factors. This research is not widely disseminated but is growing through the acceleration of knowledge in our current Information Age.

In the last days, many will travel back and forth,
and knowledge will increase. **Daniel 12:4** (NAS)

A variety of genetic factors are responsible for aging and reduction of lifespan. Although the exact mechanisms are not understood, it is obvious that the accumulated effects of all factors results in the deterioration of the information (software) of the human genome.

The human genome consists of the complete system of an individual's genetic makeup, housed in the DNA. This incorporates all the nucleotides which make up the individual "steps" on the spiraled DNA ladder and larger functional groupings known as genes and chromosomes. The genome is basically an incomprehensible set of complex instructions which directs and maintains life. The genome uses just four discrete chemicals which are analogous to four letters in our alphabet but about six billion of these letters are uniquely sequenced within the DNA molecule to form the entire code of instructions. This enormous instruction manual, the size of a dust particle, encodes for an estimated 100,000 unique proteins and numerous RNA molecules, all within each cell factory. Each of these molecules has a designated specific function and are considered to be working nano machines.

Although life has remained difficult to define, we sense that life is much more than merely physical matter. Life is the unique consequence of intelligent information incorporated within nearly every cell of an organism.

Mankind has always wanted to know from what source did all this apparently miraculous information originate? From 'Day One' he has picked up a seed and scratched his head! He knows that the information passes from one generation to another. Corn yields similar corn. A maple tree seed **somehow** produces another maple tree. This is indeed the mystery of the genome. Darwin knew nothing of the software of life and now, 150 years later, neither can science demonstrate its origin.

What then do we know? When man became disconnected from God's Eternal Life, the genetic systems which assured perfect replication of the genetic code apparently no longer function properly. Even so, scientists have identified various systems which repair and protect the DNA from copying errors and from external destructive environmental factors such as ultra violet light. Although these amazing quality control mechanisms do an excellent job, the relentless accumulation of genetic defects with each successive generation is commonly acknowledged by the scientific and medical community.

Thus the degradation of the human genome has plagued mankind from the fall of man shortly after creation. This condition can also be inferred from the record of historical longevity which is recorded in the Bible. A record of longevity and year of birth from creation is listed for the biblical patriarchs. Figure 3 is the author's plot of the biblical data. The data is from the Hebrew Masoretic text. and is also published in "The Chronology of the Old Testament" [16].

The biblical record has carefully preserved the chronology of the patriarchs so that a continuous record of genealogy is now available to us. The time span between the creation of Adam and the birth of each person listed up to Joseph is established.

Prior to the flood life spans commonly ranged between 900 and 1000 years. Lamech who lived only 777 years appears to be an exception. The Bible is silent about the cause of Lamech's death. Perhaps he died prematurely from an accident. It can also be determined from scripture that the year of the world wide flood of Noah was 1656 years after creation. Figure 4 represents typical plots of patriarchal longevity. Similar graphs can be found in various publications. Usually the horizontal axis is based upon years after Noah or years after creation. Note that the curves representing decline in Figure 4 are somewhat different from the author's. Plots similar to Sanford's are usually constructed from a mathematical function which best fits the data. The best fit curve in this case is known as an exponential curve (decline). It is noteworthy that an exponential decline is often associated with many natural phenomena such as biological decay rates.

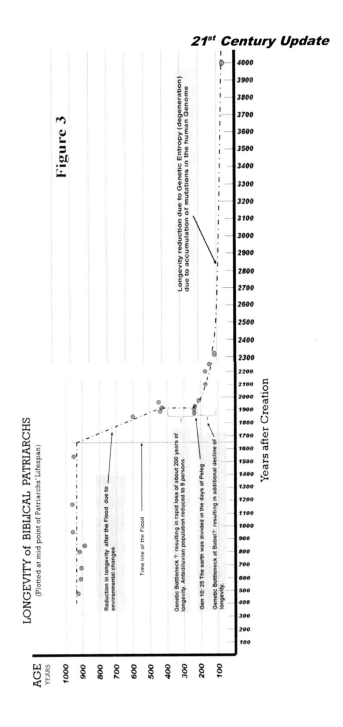

21st Century Update

Figure 3

LONGEVITY of BIBLICAL PATRIARCHS
(Plotted at mid point of Patriarchs' Lifespan)

AGE
YEARS

Years after Creation

Reduction in longevity after the Flood due to environmental changes

Time line of the Flood

Genetic Bottleneck ?: resulting in rapid loss of about 200 years of longevity. Antediluvian population reduced to 8 persons.

Gen 10: 25 The earth was divided in the days of Peleg

Genetic Bottleneck at Babel?: resulting in additional decline of longevity.

Longevity reduction due to Genetic Entropy (degeneration) due to accumulation of mutations in the human Genome

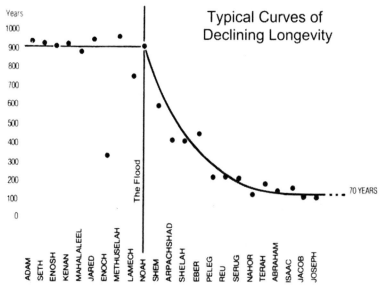

ANTE-DILUVIAN AND POST-DILUVIAN PATRIARCHS   Fig. 4

Typical Curves of
Declining Longevity

from Donald Wesley Patten, 1966, with permission

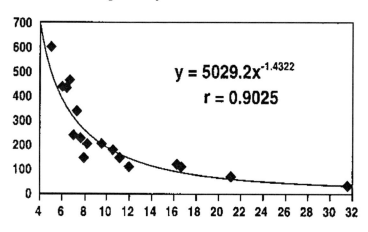

Declining Lifespan - Noah's descendants

$$y = 5029.2x^{-1.4322}$$

$$r = 0.9025$$

born centuries after Noah (lived 950 yrs)

From J.C.Sanford, 2005, with permission

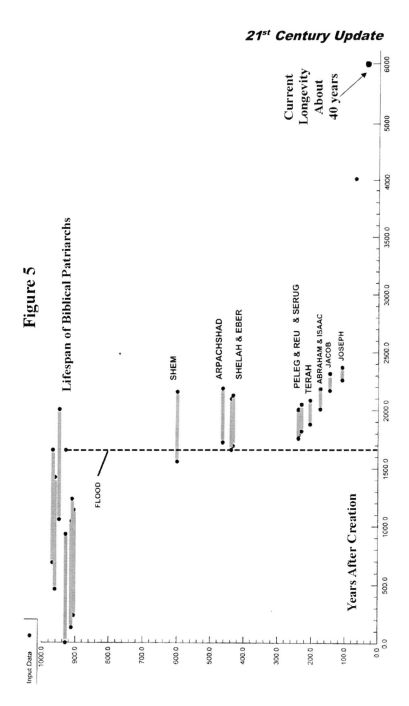

Figure 5

Lifespan of Biblical Patriarchs

21st Century Update

Current
Longevity
About
40 years

Because it is unlikely that the Hebrew scribes have fabricated this data to fit a mathematical function, the cause of the indicated rapid decline in longevity after the flood likely reflects a continuous degeneration of the human genome. This historical record is therefore consistent with the conclusions of modern genetic research which will be summarized in this briefing.

Effect Of The Flood

Now observe the author's plot of the biblical data. (Figure 3) The plot is somewhat different. I believe that this graph reveals a small but significant amount of additional information. Prior charts have been constructed by plotting longevity against the date of birth or some other date associated with the biblical patriarchs. However, this causes the plotted data points for men born after the flood to be unevenly skewed laterally to the left or right. This is because life spans after the flood reduce rapidly with time. For example refer to Figure 5: If the longevity associated with Arpachshad is plotted at his date of birth, it would be biased to the left (earlier date) relative to that plotted for Peleg. For this reason, the author believes that a more representative plot of the data would be accomplished by plotting longevity at the midpoint of each patriarch's life as opposed to his year of birth. Refer to Figure 5 which is a bar graph of the original data. Would you construct a post-flood curve connecting all points on the left side of each bar (birth date), right side or through the center of each bar?

Figure 3 represents the plot of longevity verses years after creation at the midpoint of each person's life span.

The results, although slightly different will not conform to a smooth continuous exponential curve. (or any other computer generated mathematical curve)

In any case, the characteristic post-flood decline is still obvious. It starts from the average longevity of the pre-flood patriarchs which is about 940 years. Note that the best fit to the data is linear. The straight line indicates a very rapid loss of lifespan immediately after the flood for the first three generations whose longevity quickly declined to about 410 years. (Arpachshad, Shelah, & Eber)

The next three generations experienced an even faster decline, losing another 200 years of longevity. This is plotted as a near vertical line for Peleg, Reu and Serug.

No specific cause for this drastic loss of about 700 years of life span is indicated in the Bible. But the primary factors involved may be inferred from the biblical account of the global flood, clues gleaned from geological studies and our knowledge of genetics. The two major factors are apparently due to:

1. Major environmental changes which were directly induced by the cataclysmic world wide flood.

2. Significant loss of genetic variety in the human genome.

Note that the **pre-flood** Patriarchs did not experience a continuous exponential decline in longevity even though their generations span about 1600 years. Apparently, the pre-flood environment was more conducive to longer life spans without causing any significant further reductions in longevity. Geologists can reasonably infer from examination of the sedimentary rocks that there were differences in the pre-flood flora, fauna and environment from present conditions.

It is possible that the pre-flood atmosphere had a greater sea level pressure and more uniform temperature worldwide. A reduction in atmospheric pressure after the flood would affect the ability of the body to absorb oxygen into the blood stream. Pilots who fly aircraft at high altitude know that breathing 100% oxygen may be insufficient to prevent loss of consciousness due to hypoxia (oxygen deficiency) if the air pressure is not sufficient to force O_2 molecules through the lungs for distribution into the blood circulatory system. Today increased oxygen pressure for patients is accomplished by utilizing hyperbaric chambers which are routinely used at medical centers to promote rapid healing of injured tissue. Additionally, the post-flood atmosphere may no longer provide the protection from harmful ultra violet light afforded to the pre-flood inhabitants.

Consider also that the nutritional value of food supplies may have been inhibited for similar environmental reasons. Fossil plants of the pre-flood era grew to enormous sizes. Studies have shown that the carbon dioxide content of the atmosphere can have a major effect on the growth of plants. Perhaps this is the reason why God gave permission to the flood survivors, Noah and his family, to eat meat (Gen 9:3). Apparently post-flood man now requires more protein in his diet due to environmental changes.

It is possible that the initial loss of longevity of about 500 years may be due to the degradation of the ideal pre-flood environment. However, an additional but more rapid decline in longevity follows on the graph. This may have been caused by the loss of genetic variety (traits) that was previously available in the human gene pool. The variety of potential physical traits which are available through sexual reproduction is astronomical. Nevertheless when the human population was reduced from an estimated one billion+ pre-flood inhabitants to eight, the gene pool available to the future post-flood population was greatly reduced. Perhaps this additional adverse effect appears to manifest by the loss of another 200 years of longevity with the birth of Peleg, the fourth generation after Shem. Noah's son Shem was born prior to the flood, therefore his genes were the product of the much greater pre-flood population. The loss from Eber to his son Peleg is so dramatic that as a result Eber actually outlived his son Peleg.

When the human population was reduced to just eight persons, essentially overnight, the effect upon the genome is known as a "Genetic Bottleneck". The condition occurs as a result of a significant population loss either from death or isolation.

Man's physical traits are controlled by genes. Every gene consists of two alleles, each contributed by the male and the female. Fortunately, if one of the paired alleles is impaired due to a genetically inherited mutation the other will usually mask the problem. But when a population is diminished, the variety of combinations available to the genes is likewise reduced and thus the remaining population is more prone to the negative effects of mutations.

So again, the extremely rapid linear decline in longevity after the flood could likely be caused initially by major adverse environmental changes. Then after several generations, declining longevity is also compounded by the reduction in the variety of traits available to the remaining gene pool which is imposed by a population bottleneck.

Effect of Isolated Populations

The third segment of the graph (Fig. 3) does indeed represent a smooth decline in human longevity. The relatively uniform curved line starts with the patriarch Serug. The slope of this line indicates an initial faster decline with longevity "settling" into a slow continuous reduction to present day conditions at about the time of the patriarch Joseph. The bible indicates that the earth was divided at some time during Peleg's lifetime (Gen 10:25). The prevailing interpretation of this scripture by both Jewish and church historians conclude that this reference is to the dispersion of the post-flood population at Babel.

Peleg lived for 239 years thus the exact date of the dispersion at Babel could be anytime during this timeframe. Jones[16] cites a reference to a historian (c. 250BC) who wrote that this occurred when Peleg was five years old. In any case it is apparent that another somewhat rapid loss of life span took place several generations after Peleg starting with Nahor. The rate appears to have stabilized to a consistent slow decline after the patriarch Joseph. (ref Fig.3)

Again we are faced with the cause of this accelerated loss of longevity after Serug. The author believes this condition is primarily limited to a genetic factor and is not significantly related to the environment. Geologists have obtained a comprehensive picture of conditions which ensued after the flood. Examination of the post-flood sedimentary and volcanic strata indicate that after shocks, minor crustal disturbances and mega hurricanes continued as after effects of the great flood catastrophe. In northern latitudes the climate was cooling with encroachment of the great ice sheets which eventually covered one third of the Earth's surface. These localized and temporal environmental changes were probably not responsible for this secondary accelerated loss of longevity, however.

The most probable cause was due to the imposition of a second genetic bottleneck. The confusion of languages at Babel had the effect of forcing the post flood population of dividing into small independent groups which scattered around the earth. The descendants of these **isolated** groups would again be subject to coping with another reduction of the gene pool for each group.

Current State of the Human Genome
The Genetic Problem

This security briefing has now set the stage for understanding where mankind is going. Indeed this is of interest to everyone. The last patriarch listed in the Bible is Joseph. He lived for 110 years. As you can discern, human development has not been in a constant evolutionary state of becoming more robust. The last segment of the Longevity Graph shows a continuous decline to an assumed life span of about seventy years at 4000 years after creation. Furthermore longevity is still waning! To be sure a few people have been known to live recently to as much as 120 years but these are very exceptional. The sad fact is that without the intervention of modern medical technology which most people of this planet **do not have** at their disposal, life expectancy has dwindled to a mere 40 ± years by the early 19th century in eastern Europe, western Europe and developing countries.[#17] Thus if the plot was extended to present it would further decrease to about 40 years because this is the average life expectancy in countries where sickness and disease take its "natural" course when modern medicine in not available to intervene. This state of affairs is primarily due to the relentless ongoing accumulation of genetic mutations in the human genome and the inability of our cells to duplicate perfectly indefinitely.

Notwithstanding the adverse influence of environ-mental and genetic bottlenecks, it is apparent that the quality of life as related to mankind's life span is in a continuous state of decline. The biblical data is consistent with the concept of genomic deterioration which will be elaborated in this section. This situation is, of course, completely contrary to the foundation upon which evolutionary theory is upheld. Accordingly, a corresponding evolutionary longevity plot should be ascending over time.

The deterioration of the genome is well documented in the scientific community and widely acknowledged by geneticists for the last fifty years. Below are merely a few examples, among many:

> R.A. Fisher, an evolutionary biologists and geneticists;
> ...a beneficial mutation will have only a small chance of establishing itself in a population of species and is likely to disappear. [18]

> Dr. Lee Spetner calculated the chance of survival of a Beneficial mutation to be infinitesimal and, in fact, impossible. [19]

> J.F. Crow shows that population fitness is declining because mutations are accumulating. [20]

> Dr. J.C. Sanford states "regardless of how we analyze it, the genome must clearly degenerate. This problem overrides all hope for the forward evolution of the whole genome". [21]

Contemporary evolutionists have never moved from their contention that the enormous marvelous information content of the human genome is the result of **random chance accumulation of mutations providing an outward physical benefit which is proliferated by natural selection.** In addition, it is confidently asserted that the process has been perpetually increasing in complexity and positive sophistication. Thus chemicals combined into microbes which eventually constructed the roughly 10 trillion cell complex mega-factory known as man. This seemingly simplistic principle is the absolute foundational bedrock upon which the entire evolutionary establishment of neo-Darwinian theory is built. Because there exists no viable alternative for the spontaneous or continuous creation of any organism's genome – the theory stands or falls on the validity of this principle.

Dr. John Sanford, a highly qualified geneticist, researcher labels this as the "Primary Axiom" of evolution.

> "If the Primary Axiom could be shown to be wrong, it would mean that our current understanding of the mystery of life is also wrong. This would justify a paradigm shift of the highest magnitude, and would allow us to completely re-evaluate many of the deeply entrenched concepts which frame modern thinking". [21]

Dr. Sanford is the author of over 70 scientific publications. Several of his inventions include the biolistic process known as the "gene gun", pathogen-derived resistance, and genetic immunization. He has been granted over 25 patents. Dr. Sanford has compiled substantial compelling genetic evidence which exposes the non-reality of the Primary Axiom. His conclusions are supported by the research of many geneticists and outstanding scientists.

Has modern science produced evidence to support the biblical history of declining longevity? The uncomfortable answer is assuredly "yes".

> Multiple disciplines of science
> and advanced technology of the
> Information Age
> point to the perpetual degeneration of
> the human genome.

Like every other system in this universe, the human genome has not escaped the established law of entropy. For decades microbiologists, chemists, geneticists and the medical profession at large have acknowledged that mutations have been accumulating in the genome with each generation of individuals. Sanford has aptly described this state of affairs as "Genetic Entropy".

This is no longer a debatable subject among reputable informed scientists. Multi-disciplined research has produced numerous scientific publications which verify beyond any possible credulity that the long term affect of mutations is **one way;** that is adversely degenerative. No amount of evolutionary rhetoric concerning theoretical beneficial mutations or natural selectivity can reverse this relentless trend. The absolute best that can be expected of natural selection is a slowing down of the regression, but no avenue exists for selection to halt it. This sobering conclusion, like the reality of death, is not necessarily good news whereas the human genome can not proliferate through reproduction indefinitely.

Think about this...
    If the current process of accumulating mutations

and natural selection has not been able to pre-serve the information content of the genome, then there is no reason to assume that naturalism could have created the original information package.

So who would claim that mutations are ultimately beneficial when the life expectancy of each successive generation is being reduced. Because this conclusion is disconcerting, it has been analyzed from many directions in hope of discovering a natural mechanism which would open a back door escape hatch. None is possible because, as described in a previous briefing, mutations destroy information content of the genome.

**INFORMATION IS NON-MATERIAL
THEREFORE
NO NATURAL PROCESS CAN RESTORE
LOST INFORMATION**

This leaves the Primary Axiom, the foundation bed-rock of evolution, unsupported and completely falsified. A de-teriorating genome can not be evolving but is losing ground by aging.

Recall that all life is totally dependent upon the software information/instruction package of an organism's genome. No one has been able to demonstrate the ability of any mutation to supply increased information or provide a definitive beneficial enhancement of the genome. What has been clearly made apparent, however, is that even overlooking the impossibility that such a unique condition should naturally arise; this micro benefit to the organism on a molecular level of the DNA would be completely swamped by the collateral damage of an estimated one million deleterious mutations.[#22] Geneticists know that this would unquestionably place the organism outside the arena of natural selection. Selection can only offer a species a selective advantage if its overall performance based on physical traits (Phenotype) of the organism is enhanced. In reality, a chance potential micro benefit would never surface physically due to the cumulative adverse affects of numerous genomic errors. (Sanford, p.24)

Is The Situation Hopeless?

The world's popular media and scientific community at large strive to paint a picture of a world getting better. After all, everyone wants to believe that mankind is resourceful enough to eventually solve any and all problems. However the truth is that many competent scientists have pointed out that even human intervention, including artificial regulation of selection (eugenics), cloning or genetic engineering can not completely stop the loss of information. Thus...

the human genome is time limited.

Progressive evolution is only a false hope. Science will undoubtedly prolong life somewhat in the near future but eventually the human genome will deteriorate to a critical state.

Fortunately, the Bible is clear as God has provided definite hope for **everyone** who will simply believe – trust in God's eternal Word.

> ...there is no other name under heaven that
> Has been given among men, by which we
> must be saved. **Acts 4:12** (NAS)

God has never left man without abundant provision for safety, health and general prosperity. In the long run, even death and the grave will be defeated by Jesus Christ. He is the author of the human genome and life itself.

> All things came into being by Him...
> In Him was life and the life was the
> light of men. **John 1: 3,4** (NAS)

Each and every person who recognizes this truth and receives what God has already accomplished through Him, will be reconnected to Eternal Life. The truth is usually simple and there is therefore no need to be concerned.

Society in general will try to solve its ever increasing complex political and economic problems. But the Bible prophetically assures us that the world of self government will surely degrade to a despotic state of crisis, leading to impending self destruction. At that point intervention by Jesus Christ will set up a renewed kingdom whereby the degrading effects of genetic problems will be reversed. The longevity chart will reverse its direction and climb for all those who have called on His name! Man will again live long ages and the lion will lie with the lamb.

And the wolf will dwell with the lamb,
And the leopard will lie down with the kid,
And the calf and the young lion and the fatling together **Isaiah 11:6** (NAS)

For behold, I create new heavens and a new earth; and the former things will not be remembered...

No longer will there be an infant who lives but a few days, Or an old man who does not live out his days; For a youth will die at the age of one hundred   **Isaiah 65: 17, 20** (NAS)

REFERENCES

1. Bergman, Jerry, 2000, *Why Abiogenesis Is Impossible,* Creation Research Society Quarterly, 36(4): pp.195-207

2. Denton, M., 1986, *Evolution: A Theory In Crisis,* Adler & Adler Publishers, Bethesda, MD, USA, p. 342

3. Frazer, Claire M., Jeannine Gocayne and Owen White, 1995, *The Minimal Gene Compliment of Mycoplasma genitalium,* Science 270(5235): p. 379-403

4. Roth, Ariel, 2001, Science & Origins. In Ashton, John., editor, *In Six Days.* Master Books, Green Forest, AR. p. 90

5. Behe, Michael. J,1966, *Darwin's Black Box, The Biochemical Challenge to Evolution,* New York: Free Press, Simon & Schuester

6. Zuill, Henry, 2001, Science & Origins. In Ashton, John, editor, *In Six Days.* Master Books, Green Forest, AR. pp.61-74

7. Hoyle, Fred, 1983, *The Intelligent Universe,* New York: Holt, Rinehart and Winston

8. Hoyle, Fred and Wichramasingle, 1981, *Evolution From Space: A Theory of Cosmic Creationism,* New York: Simon & Schuster

9. Coppedge, James, F., 1973, *Evolution: Possible or Impossible?,* Zondervan, Grand Rapids, MI

10. Ford, Dwain, 2001, Science & Origins. In Ashton, John., editor, *In Six Days.* Master Books, Green Forest, AR. pp. 138-142

11. Overman, Dean, 1997, *A Case Against Accident and Self Organization,* New York: Rowman & Littlefield Pub.

12. Shapiro, Robert, 1986, *Origins,* New York: Bantam Press

13. Green, David, E. and Goldberger, Robert, F., 1967, *Molecular Insights Into the Living Process,* New York: Academic Press, pp. 406-407

14. Gitt, Werner, 2001, *In The Beginning Was Information,* Christliche Literatur-Verbreitung, Bielefeld, Germany

15. Newton, Sir Isaac, 1728, The Chronology of Ancient Kingdoms Amended

16. Jones, Floyd Nolan, 2007, Chronology of the Old Testament, Master Books, Green Forest, AR

17. Schulz-Aellen, Marie, 1997, Aging and Human Longevity, Birkhauser, Boston

18. Fisher, R.A., 1958, The Genetical Theory of Natural Selection, 2ⁿᵈ Revised ED., Dover, New York, N.Y.

19. Spetner, Lee, 1997, Not By Chance! Shattering the Modern Theory of Evolution, Judaica Press, Brooklyn, N.Y.

20. Crow, J.F., 1997, The High Spontaneous Mutation Rate: Is it a Health Risk?, Proc of Nat Acad of Sci, 94: 8380-8386

21. Sanford, J.C., 2005, Genetic Entropy & The Mystery of the Genome, Elim Publishing, Lima, N.Y.

22. Gerrish, P.J. and Lenski, R., 1998, The Fate of Competing Beneficial Mutations in an Asexual Population, Genetica 102/103

No Room For Compromise

Biblical Revelation or Human Speculation

We are certainly living in the last of the biblical "last days." Jesus said we would know the season by the sign of the times. But, did you know that the Bible also accurately predicted a world wide acceptance of the current popular paradigm of the times? This was recorded about 1,940 years ago by the Apostle Peter:

> "...in the last days, mockers will come saying...
> For ever since the fathers fell asleep,
> all continues just as it was from the beginning of creation..."
> **2 Peter 3:4** (NAS)

Now, this amazing prophetic scripture declares that, in the last days, man would actually believe that all natural processes observed today have continued at the same rate with no significant differences from the beginning. When applied to the formation and geologic history of the Earth this type of thinking is called *uniformitarian theory*. It is not surprising to find that this prophecy is fulfilled today! Yes, this reasoning is actually the underlying principle that you will find in modern geology textbooks. It is often expressed in the popular media as the phrase "the present is the key to the past". The idea and its concepts seem logical because it is true that rock sediments are generally accumulating very slowly today. However, many details of the Earth's geological record provide abundant evidence to refute this long standing foundation of classic geology. Nevertheless, this **assumption** has been imagined by many scientists to extend back indefinitely in time - and thus they incorrectly conclude that the rock strata have taken millions of years to accumulate on the face of the earth. As a geologist, I believed this myself for many years!

Is this really important to know?

Well Sal, does it matter if people are convinced that the Earth is extremely old? Absolutely, because herein lies the unspoken DECEPTION which the Father of Lies has successfully planted in the minds of multitudes. **The underlying belief of the long ages of the Earth is really a sinister message.** It sounds like this:

> If the rock strata, which covers most of the earth and contains billions of fossilized re-mains, represents millions of years of death **prior** to human evolution - then death could not have been introduced by Adam. It must follow then that the creation of man and the fall of man, as declared in Genesis, is a myth... Right? **THEREFORE THERE IS NO NEED FOR THE REDEMPTION OF MAN THROUGH CHRIST!**

Even Jesus cautioned about this false reasoning when he said,

> *"If you do not believe the writings of Moses, how will you believe my words?"* **Jn 5:47**

Now what has been the inevitable conclusion of many sincere people, when the Old Testament is not viewed as reliable history or consistent with science? It follows that the New Testament naturally comes under suspicion.

When evolution and long ages of the Earth are presented as fact by the media, museums, school text books and yes, by many Christian colleges, is it any wonder that the world questions the need for the gospel message? That's right, even many in the church have yielded to this academic intimidation by trying to accommodate the proposed long ages of the Earth into scripture by inserting them between the opening verses of Genesis 1 and 2. This is commonly taught as the "Gap Theory".

WHAT DOES THE BIBLE CLEARLY SAY?

Genesis reveals to us information regarding the origin of man, his fall and need for salvation. This revelation is simply beyond the range of science to determine but the facts of science all point toward a created universe and life. The New Testament is built upon the foundation of the Old Testament and reveals Jesus Christ as the only satisfaction of that need. Thus, both Testaments have unity of purpose.

WHERE DID DEATH & THE CURSE COME FROM?

Now does scripture indicate ANYWHERE that the introduction of death and the curse, which came upon the Earth, is attributed to the rebellion of Satan? I am certain that you would need someone to help you misunderstand the clear teaching of the Bible which emphatically declares that death and calamity was introduced by **"ONE MAN"**; having entered the WORLD by Adam's sin!
Romans 5:12

Then to Adam, God said. . ."cursed is the ground because of you."
Genesis 3:17   (NAS)

Furthermore, I Corinthians 15 not only indicates that death came through Adam, but Adam is specifically identified as the first man on Earth. The scriptures know nothing of any pre-Adamic race nor do they ever attribute sin, death, or the curse, which entered the world, to the fall of Satan, prior to Adam.

**Satan definitely played a part in the Fall,
but the responsibility
falls clearly on man's shoulders**

Concerning Adam and Eve, Jesus said that they were created at "the beginning," not millions of years later, during a recreation of the Earth.

"But from the beginning of creation, male and female He made them."
Mark 10:6

Isn't it obvious when something is said to be "from the beginning" of a certain event, there must be a reasonable degree of proximity to the start involved? Now, according to scripture, Adam was created six days from "the beginning." This seems to me to be a lot more reasonably accurate than 4 billion years later, as evolutionists propose and the gap theory tries to accommodate!

THE FIRST HUMANS - WHEN?

Unquestionably, Jesus places the first human at the very dawn of creation. Other similar scriptures which involve man right at "the beginning" are: Romans 1:20, John 8:44, Luke 11:50 & 51 and Mark 13:19. Notice that all these scriptures are referring, in fact, to the beginning of the creation (Greek-Kosmos) translated as world but which literally means the entire universe (Cosmos). This coincides with the opening verse of the Bible: *"In the beginning"* God created the heavens and the earth . . . a recreated Earth is nowhere to be found in these scriptures. The opening grand verse of the Bible simply refers to the creation of the basic unformed physical components of time, space and matter (*earth*) in an unformed, unarranged state. Notice that the presence of God at the beginning also satisfies the requirement for the essential source of information which is now acknowledged in the contemporary model of our universe. In addition, many have pondered over this marvelous scripture:

*In the beginning was the Word
and the Word was with God and
the Word was God.* **John 1:1** (NAS)

By the way – what are words? Are they not the most fundamental units of coded meaningful information?

THINK ABOUT THIS …

Have you noticed that Death is a prerequisite for becoming a fossil? If the gap theory proposed by some theologians accounts for the fossils (extinct dinosaurs etc.) and the long ages for their accumulation prior to Adam, then death, struggle, and the curse had to **precede** Adam by perhaps millions of years.

BUT:

If death existed before Adam, how could - *"the wages of sin produce death"?*

Romans 6:23

If, in fact, the worldwide fossil bearing strata did represent the remains of the alleged Lucifer's flood of Gen.1:2 (which is nowhere described in scripture because pre-Adamic death would contradict scripture), then where is the evidence in the Earth for Noah's flood?   After all, Noah's flood is emphatically described in great detail in many chapters of the Bible and is spoken of as fact by Jesus, himself.

The truth is that geological evidence is coming to light, daily, to clearly indicate that the fossil bearing sedimentary rocks covering the Earth actually represent a **Huge Monument In Stone** which bears perpetual witness to the judgment of Noah's flood. This flood is referred to by the Apostle Peter several times along with his inspired prophetic prediction that man, in the last days, would eventually insist that this "Monument" merely accumulated by natural slow processes over millions of years.

"But when they maintained this (theory) they failed to recognize . . . the world at that time was destroyed by flooding (quickly)"
II Peter 3:5,6

AH - - VERY GOOD!

At the end of the six days of creation:
God saw <u>everything</u> that He had made and, behold, it was very good
Genesis 1:31

According to gap theorists, by the end of the six days of creation:
1. Satan and his angels had fallen to the Earth.
2. Earth had been destroyed in Lucifer's flood.
3. The entire pre-Adamic race, of some sort, had perished.
4. A complete world of plants and animals had experienced death and destruction and had become extinct.
   AND
5. The Earth was recreated along with Adam & Eve who were assigned to walk on a worldwide graveyard of bones (fossils)

Yet, God would survey all of this and declare . . .

"very good" ?

Proponents of the gap say, God was only referring here to the recreated Earth. But think - - even the recreated Earth would contain all of these negative conditions.

In addition, Hebrew scholars who are not necessarily Christians have stated that a straight forward reading of Genesis 1 does not support a time gap. The inspired account of creation does not necessarily require any additional scripture support for proper understanding. The first chapter will stand alone. It is uplifting to realize, however, that the Earth itself provides abundant physical evidence for the worldwide Noahic flood and a young Earth which corresponds exactly with scripture. The current state of geologic investigations attest to a catastrophic rapid accumulation of the Earth's sedimentary strata and their contained fossils.

IS CORRECT UNDERSTANDING IMPORTANT?

Now concerning the proposed gap between the first two verses of Genesis, numerous evidences exist, both scientific and scriptural, that point to a young Earth that has been inhabited by man since "the beginning." Therefore, there is no need to compromise the un-changeable account of creation in **self-defeating** attempts to harmonize scripture with the ever changing erroneous theories of fallible man, in the name of science. Such attempts have actually promoted the nonfunctional concepts of evolution and long ages and caused these anti-creation theories to tighten their grip on the media, educational systems and the unsaved of the world. As a result, Christians are faced with a more difficult task when witnessing to many sincere people who believe that the foundation of the Bible has been discredited by science. The Good News is that God promised that darkness would never overtake the light.

And the light shines in the darkness,
and the darkness did not comprehend it.
John 1:5 (NAS)

By His grace, many ministries are being raised up to help you to witness to the Truth of Creation.

I exhort you to examine very carefully what you receive as truth. Ultimately, it is the precious Holy Spirit who is our final authority and guides us into all truth as we depend on Him to do so. Let's go on now to complete our examination of the first three verses of Genesis.

At this point, God who is the prime source of information has created Time, Space and Matter initially without any specific form and uninhabited. This is logical because the raw materials of construction are not yet ready to support life. Why, there is not even any light, yet!

**Hebrew "erets" : initial basic components of matter
Biblical "dust" now understood as atoms**

VS:2 *"And darkness was over the surface of the deep
"And the Spirit of God moved over the surface of
the waters . . ."*

We are informed here that the initial shapeless mass of elements apparently was transformed into a volume of water. Water appears to be an important building material from which the entire universe was created.

It is now the moment for the Holy Spirit to be directly involved in creation. The word "moved" is also translated as 'shake', 'stirred', 'moving to and fro' or 'flutter' but I can assure you this is no common fluttering or hovering. The Hebrew implies a rapid back and forth motion so it has been often assumed to be analogous to the rapid wing movements of a bird. Perhaps the best word available today to describe this rapid back and forth (oscillatory) motion is "vibrating". Is this coming together? Remember, nearly all forms of energy can be described in terms of vibrations. I believe we are given here, in VS: 2 **God's revelation of the initial source of all energy** which fills the universe. It is described here as the impartation of energy (vibrations) by the Spirit of God, Himself upon the initial dust (atoms) of created matter.

> ➤ **Prov 8:26** refers to Jesus existing from eternity:
> even <u>"before the first dust of the world"</u>

The energy imparted to the elements of Hydrogen and Oxygen provided the initial bonding energy to form water molecules. Water molecules are also known to have a weak attractive force known as cohesion. This attraction between molecules would cause the mass of water to coalesce into a huge sphere, thus beginning the initial stage of creation of planet Earth.

*God set a circle upon the **surface of the deep*** Proverbs 8: 27

Water sphere made in the International Space Station

⟶ 3 inches ± ⟵

Cohesion due to attractive forces between H₂O molecules

Jesus tells us, in Acts, Chapter 1, that you will receive POWER when the Holy Spirit comes upon you. In fact, the Bible is replete with scriptures referring to the Holy Spirit as the ultimate source of power or anointing of God.

Does this correspond with our scientific observations today? Absolutely! In fact, one of the most fundamental and highly utilized laws of Physics states that all the energy in the universe is a *fixed* quantity that cannot NOW be created or destroyed. This is known as the First Law of Thermodynamics. The awesome implication of this known physical law is that energy could not have created itself a little bit at a time through some evolutionary process, therefore, you are back to two choices, again:

1. Eternal Energy
OR
2. Eternal God (All Energy created suddenly)

If you choose to reject Choice #2, you must recognize that, as with the **origin of matter**, science is again mute when faced with explaining the **origin of energy**. But we know its source by revelation... the Holy Spirit.

Where are we now, anyway? Well, from the information presented, I think you can perceive that creation has been a continuous process, up to this verse. God, the Father - The Ancient Of Days - has been the architect of all this. The Holy Spirit imparts energy to the formless raw materials, thereby completing the three basic physical components of the entire universe.

TIME - SPACE - MATTER ⇄ ENERGY

From here, the stage is set for these components to be synthesized into the current form of the ordered universe.

The last verse we will examine is:
VS:3 *"Then God said . . ."*

It is definitely significant that the **first words** of God recorded in the Bible are revealed to us at this point in time, in the process of creation. Who is the Word Of God? Jesus Christ, of course!

*"In the beginning was the **Word**....and the **Word** was God"*
John 1:1 (NAS)

183

> *"All things came into being **by Him** and apart from Him*
> *nothing came into being that has come into being."*
> **John 1:3** (NAS)

Before we go any further, let's realize that right here in verse 3, at the very opening of scripture, God is revealing to us the mechanism behind all creative power.

God's Word

Yes, it is His word that is obviously the original source of all the directive information that not only created but also continues to maintain this universe (Heb 1:3). And it is this information, recently discovered, which is still directing the activities of every living cell.

The good news is that the New Testament tells us that **God did not choose to limit His creative words to just Himself.** But by His grace, He has made His words (the Bible) available to every Born Again, Spirit Filled believer. In addition, we are encouraged to speak His word into the circumstances of our lives in order to help build His Kingdom on the Earth. Isaiah prophesied of this:

> *"I have put My words in your mouth and have*
> *covered you with the shadow of My hand, to*
> *establish the heavens, to found the earth,*
> *and to say to Zion, You are My people."*
> **Isaiah 51:16** (NAS)

So let's continue on with **VS:3** "Let there be light. . ."

Right here is where the skeptic will call a "time-out!" With typical human reason, it has been legitimately questioned "How can light exist, when the Bible clearly states that the sun was not created until day 4?" Isn't this a contradiction and a scientific impossibility? Well, first of all, none of the creation events have a scientific explanation for nothing has ever been literally created by the current physical laws now in operation. Scientists have merely learned, to an extent, to manipulate the matter and energy that now exists as a result of creation through the clever inventions called machines. But, because this is an important question, many theologians and ministers have attempted to provide a natural explanation for the "first light" of VS 3. Invariably, those attempts have fallen far short of any scientific credibility and may have, on occasion, even provoked ridicule of the Christian community.

What, Then, Is The Source Of THIS First Light?

I seriously doubt if anyone will ever be able to explain a physical mechanism for this phenomenon, because I believe its source is supernatural. This event, the bringing forth of the first light, like the former creative acts was provided by God for a very significant purpose. I submit to you that scripture reveals the existence of the first light which flooded the universe, devoid of a physical source, to IMPRESS upon mankind the fact that:

God - The Majestic Glory- Is of Himself
The Primary Source Of
ALL LIGHT AND ALL LIFE!

We are to understand that all sources of light now existing in the universe are merely secondary physical sources, which initially obtained their energy from **HIM.**

"God Is Light"
I John 1:5 (NAS)

Recall Revelations 21: (NAS)
> "And I saw the Holy City, the new Jerusalem
> coming down out of heaven from God **AND THE**
> **CITY HAS NO NEED FOR THE SUN OR THE MOON."**

So, we can clearly determine, from scripture, that we do not necessarily need the Sun to produce light!

Revelations 21: continues
> "For the Glory of God has illuminated it
> **AND ITS LAMP IS THE LAMB"**

I submit to whoever will receive - that we have here, in VS:3 of Genesis, None Other than the glorious Lord - The Word Of God - fulfilling His part in creation. His entry brilliantly illuminates the globe of initial matter as His spoken words begin to organize the Earth and the entire universe. (Heb 1:4).

Of Jesus, **THE FIRST LIGHT**, the scriptures refer in this manner:

"The origin of the creation of God" **Revelations 3:14;**
"The God who commanded light to shine out of darkness"
2 Corinthians 4:6

"I am the Light of the world;
he who follows Me shall not walk in the darkness
but shall have the light of life" **John 8:12**

Now I ask the reader of this book - Do You Know Him?

Again, the purpose of VS:3

"His light dispels <u>All</u> darkness"   **John 1:5**

God is impressing upon us here that there should never be a need for any nation, tribe, or individual to ever worship the sun, moon, or stars (Deut. 4:19).

Why, then, you might ask, did Jesus command His own glorious light to illuminate the creation?

Now, please be cognizant of this –
Science is LIMITED to that which is:
>Directly Observable
>Reproducible
>Testable

No amount of scientific observations . . .

No depth of human intellectual reasoning . . .

can ever enlighten man concerning the **ultimate origin** of life and matter, which could never be observed because NO ONE was there. Neither could the origin of life and matter ever be reproduced by any natural process because creation was a one time event - NOW COMPLETED!

THEREFORE

God's own glorious illumination was given to us by His grace to emphasize that the **FIRST LIGHT OF MAN'S REVELATION** is literally Genesis, Chapter 1. And this gift of enlightenment was given to us as a supernatural **Gift from God.** Otherwise mankind would always have remained in darkness concerning origins, being limited only to his own foolish speculations.

Consider this scripture carefully: (referring to Genesis 1:3)

For God, who said, 'Light shall shine out of darkness'

is the One who has shown in our hearts to give
THE LIGHT OF KNOWLEDGE *of the glory of God,*
in the face of Christ."   **2 Corinthians 4:6** (NAS)

No - the Genesis account was not given to prove that God exists. But to reveal to us as a "first light" something that could **NEVER** be determined by man's own research or intellect ...

THE ORIGIN AND PURPOSE OF LIFE ITSELF

"there was the true light which coming into the world, enlightens every man"   **John 1: 9** (NAS)

To those who receive this revelation, there is no inconsistency with observations of our physical world. Once a person realizes the integrity of Genesis, then the need for Jesus, the Messiah, becomes evident. The Bible does indeed provide reasonable answers to sincere questions regarding origins in agreement with the physical sciences. For those who have personally received the Author Of Life, your knowledgeable witness should be a demonstration of the Holy Spirit's power (1 Corinthians 2:24) because we must ultimately rely on the Holy Spirit working through us to *"convict the world concerning sin, righteousness and judgment"* (**John 16:8)**.

It is the goal of this book to reveal to you the ultimate truth and absolute integrity of God's Word from the very first page of the Bible.

It's Time For A Decision

By NOW it should be obvious that every person is continuously exercising his faith when choosing what you believe by an act of your will. You can either accept the revelation of scripture and the collaborating witness of "all creation" and believe in the Eternal God of the Bible - Jesus Christ

OR

by default, you **are placing your faith** in the eternal existence of matter by believing that TIME, CHANCE and MATTER organized and programmed **itself** into the entire ordered universe and all living organisms. If you choose the latter, you must believe in evolution in spite of the fact that no known physical or biologic mechanism for this has ever been observed, demonstrated or verified by the scientific community. In fact, as previously exposed herein, the theory of evolution **violates** numerous established laws of physics, chemistry and genetics.

Multitudes of people worldwide, young and old, are awakening to the truth. The brief testimony that follows is typical of the many new decisions being made every day. Mr. Richard W. Stevens was a computer science major at the University of San Diego. He now resides in Phoenix, Arizona and practices law as an attorney.

"Nine o'clock one Sunday night in 1968, a prophecy-oriented evangelical program came on the radio. I remember it vividly. It changed my life. But that program was not about the Gospel or the End Times. It was about Darwinian Evolution. And for the first time I heard someone very credibly question and ridicule that theory on its own terms.

Raised in a mostly un-churched home that valued reason over superstition, I thought that science was the only way to truth. My family and I did not dislike or even reject Christianity; we just thought it was a belief system like other religions and totally separate from the real world. It was comforting to some, but not a way to understand the universe. Evolution, being the proven explanation for life on Earth, was another grand triumph of science. Genesis Chapter One had to be allegory or myth; it had no other relevance to me.

The radio minister, however, asked me to think about how a reptile evolved into a bird. The first lizard in the sequence is perfectly healthy, but her offspring has a **mutation**. A pre-wing like structure protrudes from both sides of the young lizard. That lizard must drag the structure around with him. It slows him down when trying to escape predators and when chasing after food. All the other lizards get their prey, but the poor mutant is still huffing and puffing as he clambers to grasp the elusive grasshopper.

During mating season, the other lizards can get a date with compatible normal lizards. The poor mutant lizard's mating dance is awkward and he doesn't even look much like healthy reptile husband material. Assuming that he has escaped enough predators and managed to capture enough food to survive, his chances of reproducing are less than other lizards. All in all, under **natural selection** this mutant stands a lowered chance of individual survival, let alone making more individuals like himself.

The poor mutant cannot even fly with his silly fleshy appendage. Jumping off a rock or tree, he doesn't glide and soar; he plummets. He's just a loser among lizards, and a lousy excuse for a bird.

This simple word picture of the evolution of birds forced me to realize that evolution had to do more than **"suppose this"** and **"imagine that"** if it were to be considered fact. Because I never found a plausible evolutionary explanation for changes in species, I was for years an agnostic who rejected evolution.

I realized that if evolution is untrue, than science can- not be trusted in everything it declares. If evolution is untrue, then the existence of a creator becomes possible, even necessary. If a creator is necessary to explain life on Earth, then Genesis Chapter One just might be true.

When I learned how basic Christianity elevates and liberates mankind while reconciling us with God, and that Jesus' resurrection was factually likely to have actually occurred, then there were no remaining intellectual barriers to belief. The Holy Spirit needed only to nudge me into a faithful relationship with God and Christ".

Friends, the only reasonable explanation of life is life itself. Since "Day One" man has continuously observed that life

ONLY comes from life. This leads any honest person to the plausible conclusion that life is a product of intelligent design and purpose.

Although man corrupted this wonderful original creation by his own decision to walk separate from his creator, yet the love and mercy of God reached out and provided **complete** restoration through Jesus Christ. The scriptures are clear - no one can set himself right before God by his **own good works** and intentions. To try and do so is only a futile exercise in religious activities. Receiving the **free** gift of Jesus Christ runs contrary to our pride because it requires us to accept that a right standing with God was purchased for us through the merits of Jesus Christ. If you receive the truth that God is the creator of all things and, in Christ, He paid the penalty HIMSELF for all our shortcomings; then you are **assured** that God will restore your fellowship with Him and receive you as a son (or daughter) - no questions asked.

This new birth is not religious ceremony like confirmation, sacraments, church membership or water baptism. Jesus told a very intellectual, well respected Jew named Nicodemus -

"You MUST be born again" **John 3:7** (NAS)

You see, the issues of Eternal God or Eternal matter are of utmost importance because they will impact your ETERNAL WELFARE. **Therefore, you want to get it right!**

See to it that no one takes you captive through philosophy and empty deception, according to the tradition of men, according to the elementary principles

of the world, rather than according to Christ
Colossians 2:8 (NAS)

> **Do not permit the unfounded theories of men
> to mislead you in
> ETERNAL MATTERS**

Go to the Bible **yourself** and make certain within your own heart that you are right with God. Humanism, intellectualism or even church membership cannot take the place of the new birth by Christ, the Creator. You are either restored to the **image of God** through the spiritual new birth or you are not.

Your eternal life can start right now, as you sincerely say to Him...

> Lord, I believe you are the creator of all life and
> I believe you chose to die for me in my place.
> You rose from the dead to provide me with a new
> life of eternal fellowship with you.
> I thank you for the **free** gift through the merits
> of Jesus Christ. AMEN - so be it

Now, if you have sincerely prayed in this manner for the first time, it is important that you tell a Christian friend. The Bible says:

> *If you confess with your mouth, Jesus as Lord, and
> believe in your heart that God raised Him from the
> dead, you shall be saved.*

> *For with the heart man believes, resulting in
> restored fellowship with God, and with his mouth
> **he tells someone**, resulting in salvation.*
>   **Romans 10:9,10**

194

CREATION SCIENCE ON THE WEB

Would you like to understand more about science, the Bible, creation and origins? I predict you will find Creation/Evolution issues coming to the heated forefront of controversy in the near future. The public school system and media have been teaching evolution as a fact for decades. As children realize that what they are taught is contrary to scripture, many parents do not know how to adequately answer their questions. An abundance of information is available free to provide you with answers to perplexing questions regarding the true history of the Earth, age of the Earth, the fossil record, ape-men, dinosaurs and the world wide flood. Learn how all these relate to the Bible. This information is available on the worldwide web.

CHECK OUT THE FOLLOWING EXCELLENT WEB SITES:

www.AnswersInGenesis.org
www.ChristianAnswers.net
www.Bible.com
www.ICR.org
www.CreationEvidence.org
www.CreationResearch.org
www.Microscopy.org
www.CreationSafaris.com
www.DrDino.com
www.CreationMinistries.org

SUGGESTED ADDITIONAL READING
(available from author's website)

The Answers Book, Master Books

The Genesis Flood, Whitcomb & Morris, Institute For Creation Research

The Young Earth, John Morris, Master Books

In Six Days – Why 50 Scientists Choose To Believe In Creation, Master Books

Darwin's Black Box, Michael Behe

Answers (magazine), Answers In Genesis.org

Not By Chance, Lee Spetner, Judaica Press

Evolution: A Theory In Crisis, Michael Denton, Adler & Adler Publishers

The Biotic Message, James ReMine, St. Paul Science

Evolution: The Fossils Still Say No!, Duane Gish, ICR

In The Beginning Was Information, Werner Gitt

Origin By Design, Harold Coffin

Thermodynamics and The Development Of Order, Creation Research Society

Genetic Entropy & The Mystery of the Genome, Dr. J. C. Sanford, Elam Publishing

Thousands not Billions, Dr. Don DeYoung, Master Books

Living Fossils, Dr. Carl Werner, New Leaf Press

Free Resource:

http://www.azosa.org/resources/powerpoint

Creation Science Power Point Presentations and slides available on the web for download. To be used in presentations, home schooling, or in slide rotations prior to church services. These are a way to introduce Creation Science and the truth of Scripture to individuals and congregations in a simple, non-challenging and inform-ative way.

Biblical Creation Mega-Store
http://www.godordirt.com/bookstore.htm

CPSIA information can be obtained at www.ICGtesting.com
263301BV00002B/6/P